Paul Goodwin is the author of *Forewarned: A Sceptic's Guide to Prediction*. A professor at the University of Bath and a former adviser to government departments, he has taught special courses on statistics for trade unionists, art curators, surgeons, actuaries, civil servants, CEOs and sixth-formers. It might not sound like much, but that's actually a lot of real-world experience for a mathematician.

Something Doesn't Add Up

Surviving Statistics in a Number-Mad World

Paul Goodwin

P

PROFILE BOOKS

This paperback edition first published in 2021

First published in Great Britain in 2020 by
Profile Books Ltd
29 Cloth Fair
London
ECIA 7JQ

www.profilebooks.com

1 3 5 7 9 10 8 6 4 2

Typeset in Dante by MacGuru Ltd
Printed and bound in Great Britain by
CPI Group (UK) Ltd, Croydon, CRO 4YY

A CIP catalogue record for this book is available from the British Library.

ISBN 978 1 78816 259 3
eISBN 978 1 78283 549 3
Audio ISBN 978 1 78283 680 3

To my wife, Chris

Contents

Introduction

Statistics will solve all our problems

In the early 1970s I got a job as a statistics lecturer at Hull College of Commerce. I was motivated by a passionate belief that teaching people statistics would help to make the world a better place and I used to repeat H. G. Wells's famous quote whenever I encountered doubters: 'Statistical thinking will one day be as necessary for efficient citizenship as the ability to read and write.' I'd even rote learned the quote for the job interview and managed to recite it word perfectly in answer to the College Principal's last question. I was convinced this had impressed the panel and got me the job.

With the promised arrival of huge computing power, I was sure that statistics would bring light to dark places. It would reveal hidden patterns, linking lifestyle factors to diseases and detecting impediments to productivity in Britain's desultory industries. It would allow people to make better decisions either at home or at work and it would serve democracy by enabling them to distinguish the mumbo-jumbo from the facts.

I also found an appealing certainty in statistical calculations. You were either right or wrong. I'd read George Orwell's *Animal Farm* when I was much younger and had enjoyed the tale. Years later, someone told me it was an allegory of the Russian Revolution. That worried me. I'd read the whole novella and missed the point. At least J. R. R. Tolkien told readers that his *Lord of the Rings* was not intended to be allegorical. Otherwise, how did you know? I would watch a film in the cinema only to walk home with friends who would fiercely debate what the film was about – lost innocence perhaps, or alienation, or existential angst.

Sometimes it seemed that you could twist the book or film to mean whatever you wanted it to mean. Maybe that was the idea of art, but it would have been so comforting to have known what the correct answer was. Of course, it is often said that you can twist statistics to prove anything you want. But, in those days, I thought this could only result from the deliberate misuse of statistics – hiding numbers that were inconvenient, using misleading scales on graphs, or repeatedly carrying out small-sample surveys until, by chance, one gave you the result you needed. The honest use of statistical methods, I believed, would lead you to a truth that was indisputable. At least there was no doubt that 6 plus 8 equals 14 or that the mean of 2 and 7 was 4.5.

My attraction to statistics was accompanied by some rapid and exciting developments. We take them for granted now, but at the time they seemed amazing. By the mid-1970s electronic calculators were becoming widely available. People showed me with glee that if you typed 71011345 and turned the calculator upside down, it spelled Shell Oil. Esso Oil was 7100553. Then, as the years went by, it was astonishing to see what computers could do with numbers. With a little tuition you could write Basic programs that whizzed through daunting calculations in seconds. Statistics now came without the tedium of adding up columns or multiplying dozens of numbers on paper. Before computers you had to think hard about what you wanted to measure, and why, because obtaining a result took so much time and effort. You had to be parsimonious and frame your questions tightly. Now there was a temptation to let the computer loose on data, with little guidance, to see what odd things it could uncover. It could show you in an instant whether safer playground equipment was associated with childhood obesity or whether the number of rockets launched into space per year worldwide was correlated with the number of sociology degrees awarded in the USA. Of course, it wouldn't tell you whether these results were useful or made sense, but we humans are brilliant at inventing explanations for surprising correlations, even if they are, in reality, spurious.[1]

It seemed as if nothing was real unless you could measure it. I saw a Gallup report[2] in 1981 on satisfaction with life in different countries that showed Northern Ireland scored 7.68 out of 10, while the rest of the UK only mustered 7.67 – and that was despite the Troubles Northern Ireland was suffering at the time. Denmark came top of the list and Japan, bottom. Here was measurement to two decimal places. It looked scientific. So a nation's satisfaction with life did seem to be a real thing. But what about morality? What about love? Perhaps you could measure a person's heartbeat or their perspiration level when they were told to think about their partner. Combine these measures and you might find that their professed love actually fell below the 'true love' threshold of 5.0. Or perhaps love didn't really exist at all. There are two worlds, a highly mathematical colleague once told me – the world of mathematics and the world of waffle. Love clearly belonged to the latter.

Numbers on the march

Fast forward four decades and numbers are triumphant. Datafication involves transforming all aspects of our lives into data. How we sleep, shop, exercise or communicate with friends on social media – even our personalities or how we sit in a car or clean our teeth – can now be represented by numbers. And all these numbers can be swept up and devoured by that behemoth of our time, Big Data, allowing secret algorithms to build models of who we are and what makes us tick.

But that's not all. When it comes to choosing where to study, where to live or where to seek medical treatment, our decisions are guided by league tables that rank cities, universities, schools, hotels, restaurants and hospitals using precise scores measured to one, two or even three decimal places. Sometimes it feels as if we are staggering around in a fog of numbers. Like octopuses, politicians instinctively squirt protective clouds of statistics around themselves when under attack. Hopelessly confused, many of us are forced give up on the 'facts', so we vote for the candidate who is 'the guy you'd want to have a beer with'.[3] Or we are inspired to support the candidate who shakes the most with angry passion when giving their speeches, even though something deep down tells us that what they're saying doesn't quite seem plausible.

In many areas of our lives numbers have become our master – and an unforgiving one too. Serried ranks of workers in call centres are measured on just about every move they make. One worker I know received a message complaining that she had returned from her coffee break 18 seconds late. Percentage of calls answered or missed, average time spent talking, percentage of callers who hung up – they lend themselves to pretty graphs on managers' screens, but they can be bad news for the worker who has the longest or shortest bar on the chart, depending on what is being measured. Even university lecturers now fear the cursed fate of numerical underperformance. In one scheme, those whose average feedback score from students was less than 3.5 out of 5 were hauled before a committee to explain their failings. In the UK, university researchers are admired if they are 'sixteen-pointers', which means that, in a relevant period, they have four papers in journals rated at the top level of 4. In some institutions you need to look out if you are a mere 'eleven-pointer' or less.

More frightening is the development in China of the Social Credit System, which will be mandatory there by 2020. A single number will measure the 'trustworthiness' of each citizen based on factors like promptness of bill payments, shopping habits, messages they send, their choice of online friends – and even what these friends say and do. Buying video games, for example, is likely to lower a person's score – it suggests indolence. Buying nappies will raise the score as it is likely to be associated with a sense of responsibility as a parent. Those with low scores will experience a whole range of disadvantages, such as slower internet speeds and restricted access to jobs, restaurants, leisure facilities and entertainment venues. High scores will bring greater freedom to travel internationally and even access to the VIP check-in at airports.[4]

Sometimes the numbers that deem what's acceptable or not acceptable, or what's good or bad, can seem very strange. In 2017, the World Bank's chief economist, Paul Romer, aimed to encourage clear communication by insisting that no flagship report would be published if the word 'and' exceeded 2.6 per cent of the total.[5] At least in this case those under the numbers' lash rebelled. Subsequently, Romer was relieved of his management responsibilities.

It's more common for people to react to numerical targets by adapting their behaviour so that they achieve them even when this makes no sense at all. There are uncorroborated stories of shoe factories in the Soviet era that found they could more easily meet their production targets if they only produced shoes for left feet.[6] In 1999, the 6.20 a.m. Connex commuter train service from Ramsgate in Kent failed to stop at three scheduled stations on its way into London, allegedly to ensure that the company could meet a 100 per cent punctuality target.[7] And, in one week in April 2013, there was an astonishing 191 per cent rise in demand for theatre tickets in London. Apparently, the BBC's Class Calculator had just gone online, allowing people to determine their social standing. Your rating was enhanced if you could say that you 'go to the theatre'.[8] In the Chinese Social Credit System your score – and those of your friends – will rise if you send online messages that say positive things about the government and the economy.

Some of us have become slaves to numbers. My fitness tracker just vibrated to warn me that I have not walked the required 250 steps in the last hour. To meet my daily target of 10,000 steps I've sometimes been forced outside into raging storms at 11.00 p.m., otherwise I'd face a sleepless night through guilt. If I missed the target, I would only have myself to blame for all the ills that come through indolence. Even if I've achieved 9800 steps, that could be the thin end of the wedge – the start of a slide into a life of sedentary indulgence.

Now the tone of each day is dictated by numbers. It's a bad day when I've gained a pound on the scales in the morning, when my sleep monitor indicates that my minutes of REM sleep fell short of the normal range for men of my age, when the Amazon sales rank of my last book has gone down to 23,707, three thousand lower than when I last looked – ten minutes ago – and when a research-monitoring site tells me that only 23 people read my papers last week (a decline of 24.8 per cent on the previous week).

It's not surprising that there's a new phenomenon called social media depression, which, at least in part, is caused by the presentation of numbers on sites like Facebook. If other people have more friends, more likes for their photographs, and more comments on their posts, they must, some users reason, be more likeable and interesting. And the resulting envy can lead to depression, particularly among teenagers. The fact that many users have never met, and have little knowledge of, their hundreds of 'friends' is forgotten. One user wrote about how he cried himself to sleep because only 36 out of his 600-plus 'friends' had wished him a happy birthday.[9]

Two worlds

I now realise that my mathematical colleague who argued that there are two worlds – mathematics and waffle – was wrong. There *are* two worlds, but they are the world of reality and the world of numbers. And the second world is usually at best a simplification of the first and at worst a gross distortion of it.

Simplification can be useful. Brilliant though our brains are, they have limited processing capacity, so numbers can direct our attention to the guts of problems and the essence of what is going on – free from the fog of detail. When the *Washington Post* reports that President Trump made 1628 misleading claims in his first 298 days in office,[10] that's a clear message that something may be amiss – though of course we then need to look at the evidence underlying the report. When the *Daily Telegraph*[11] reports that in the UK women earn, on average, 9.4 per cent less than men, it cuts through the detail that there are some very highly-paid women and low-paid men, that there are regional differences in pay, and that pay will depend on educational background and age. Again, we would need to check how the average was calculated.

Numbers can also help us to gain clearer insights into other people's thinking. For example, if a doctor tells me that there's an 80 per cent chance that a forthcoming operation will be successful this might give me a better idea of her thinking than a vague statement like 'it's quite likely that the operation will be success'. That, of course, assumes that I am comfortable thinking in terms of probabilities, but forcing the doctor to quantify her estimate might make her think harder and help her to clarify her own thinking.

But often, when you are asked to put a number on something – for example, when responding to an interviewer in the street – you will realise that your response is an oversimplification. How satisfied with life are you on a 1 (extremely unhappy) to 10 (extremely happy) scale? Your response might depend on the weather, whether you've just had a coffee, whether a shop assistant just snubbed you, or a subconscious desire to convince the interviewer that you are not a miserable git. Even discounting these factors, can you really condense the complexity and diversity of experiencing life into a single number? It might be a rough reflection of your general state of happiness, but surely much of the richness of life is lost in that impoverished number. The key is to exploit the simplicity of numbers as far as this is useful, but also to be aware of their limitations and to realise that they usually only tell you part of the story.

We should be particularly concerned when the world of numbers is a serious distortion of the reality it aims to represent. Distortion can be deliberate – marketeers, politicians and campaigners often exploit a range of numerical conjuring tricks to sell us untruths. Newspapers sometimes aim to surprise us with dubious figures to boost their circulation. We are told that 80 per cent of Britons would be happy to emigrate to Russia if President Putin gave them free land in the Siberian wildnerness.[12] Really? But distortion can also result from ignorance or carelessness. Flawed surveys can drive political decisions. Scientists sometimes draw important conclusions without reflecting on the rationale of the statistical methods they employed.

The problem is that, for many people, numbers are uninteresting, untrustworthy, mysterious or indigestible. Whatever the cause – culture, an intimidating maths teacher when we were at school, or the perceived mendacity of the number merchants – the result can be either a passive acceptance of misleading numbers or a refusal to accept useful numbers. In the latter case, this can encourage a search for the truth in unreliable qualitative evidence. A newspaper story accusing a family of being scroungers eclipses statistics on welfare claimants actively seeking employment. A photograph of a pensioner attacked by thugs supplants reassuring statistics on the rarity of crimes against the elderly.

Even those who are educated in mathematics and those who like numbers are not immune. Psychologists have shown that we have two systems for thinking – a fast, effortless and intuitive system known as System 1, and a slow, deliberative and analytic system, known as System 2.[13] The latter requires effort and hard work, so we often default to the easy intuitive approach. But this can catch us out. I recall a colleague who had a PhD in statistics celebrating that her syndicate had won a significant sum of money on the National Lottery. 'We'd better change our numbers before we enter next week's draw,' she said, reasoning that it would almost be impossible for the same set of numbers to come up two weeks in a row. System 2 thinking would have told her that she was implying that the lottery balls had a memory and would deliberately be organising themselves so that a new set of numbers would appear in the next draw.

All of this limits people's ability to be effective citizens, achieving the opposite of what H. G. Wells hoped for. Indeed, some people argue that this cursory relationship with numbers is a major threat to democracy. During the Brexit referendum in the UK, a bus toured the country emblazoned with a message that leaving the EU would allow an extra £350 million a week to be spent on the National Health Service. Evidence from the UK Statistics Authority that the figure was a lie apparently had little impact, as did a barrage of depressing statistics on the effect of a Leave vote on employment, wages, the cost of living and inward investment.

In contrast, rather than being sceptical, there are those who embrace numbers to excess, believing that there is nothing beyond them. They may also be living in a world of falsehoods. The number of steps my tracker records each day is not the same as fitness. To the Facebook user, the number of friends their smartphone displays is not the same as popularity or sociability. To an unemployed person in North East England, a 3 per cent growth in GDP being celebrated by the government is not increased wealth. Perhaps these number lovers are just as easy to deceive, so their ability to be effective citizens in a democracy is also impeded.

This book argues that we are often too accepting of numbers when they mislead us, while we ignore them when they have important things to tell us. It shows how both an aversion to numbers and an excessive zeal for them can blind us to the truth – a truth already under challenge in the so-called 'post-truth' era – and what we can do about it. It aims to help find the sweet spot between the dangers posed by the extremes of number resistance and unquestioning acceptance so that we can exploit numbers where they bring benefits, but also appreciate their limitations and challenge them where they mislead. Armed with that skill, we can all make better decisions.

So how can we get people to interact more insightfully with numbers? To risk a cliché – it's all about striking a balance. At the end of the book we will consider how this might be achieved by making numbers more palatable and engaging. You'll also find a toolkit there that aims to help you to judge the validity of a statistic. Above all, we need to control numbers rather than allowing them to rule our lives. This isn't easy. I'm trying to avoid the temptation to discover the Flesch Reading Ease score for this Introduction. After all, can a single number really tell you how readable a book is? I shall resist.

OK, it's 52.9. That makes me feel much better.

1

Rank Obsession

Best of the best

When I put the search term 'of the year award' into Google, it found 6,840,000 results. There was Museum of the Year, Business Analyst of the Year, MP of the Year, Pension Scheme of the Year, Sermon of the Year, Grave Digger of the Year, Coffin Supplier of the Year, and multiple awards for Book of the Year, Player of the Year and Employee of the Year. There were even awards for Loo of the Year and one for the year's Oddest Book Title[1] – for which books titled *Nipples on My Knee* and *Renniks Australian Pre-Decimal and Decimal Coin Errors: The Premier Guide for Australian Pre-Decimal and Decimal Coin Errors* were hot contenders in 2017.

It's surprising how competitive people can be when there's a potential title on offer. There have been many reports of gardeners sabotaging their rival's prize plants in the dead of night,[2] and as twitchers battle to dominate the rankings, the supposedly gentle hobby of birdwatching in Britain has been described (by Anthony Faiola in the *Washington Post*) as 'truly savage'.[3] In some cases, awards provide an incentive for the delivery of improved services or products and they bring satisfaction and pride to the winners, though this may be countered by the collective demotivation and resentment of the many candidates who didn't win.

Besides the elation or misery they may bring to the contenders, rankings and awards can have a significant influence on our decisions. We might choose to study at a university that's been crowned University of the Year, buy a hatchback that's Car of the Year, invest our savings on the say so of a Financial Advisor of the Year, or promote an academic who's won the Journal Paper of the Year award. So does choosing *numero uno* really mean we are getting the best of the best?

Sometimes ranking is a matter of judging whether an apple is better than an orange. And if you can't tell, then toss a coin. The award's in place and someone has to win it, otherwise we'll look foolish. Many years ago a relative of mine was asked at the last minute by a neighbour to join him as a judge in a competition of dance troupes of teenage girls. One of the regular judges was indisposed. 'But I know nothing at all about dancing,' my relative protested. 'Don't worry, when you get the ranking form, just write anything down. That's what I do,' chuckled the neighbour. Somewhere, a middle-aged woman might be fondly recalling the day she was in a group that won Dance Troupe of the Year award and proudly showing photographs to her grandchildren.

Even if we try to do an honest job of judging ranks, we are likely to be stumped when there are lots of candidates for the top place. Newspapers and magazines regularly have columns telling us the best places to eat in London, the best cars of the year, or the best new books to read. Of course, strictly speaking, these claims can only be accurate if *every* candidate has been thoroughly assessed and compared. Have people writing these columns eaten in every restaurant in London (there may be at least 17,000 of these), driven every brand of new car launched in the past year (at least 67 were expected in the last nine months of 2018 alone), and read every new book published (an estimated 173,000 titles were published in the UK in 2015)?[4]

Identifying the best of the best is likely to be particularly important to people who are what the American psychologist Barry Schwartz calls 'maximisers'. Maximisers spend much of their time in an unrelenting search for a better job, a better partner, a better car or a nicer place to live. One question on Schwartz's diagnostic questionnaire asks whether the respondent agrees with the statement: 'I treat relationships like clothing: I expect to try a lot on before finding the perfect fit.' Another asks to what extent the respondent sees themselves as 'a big fan of lists that attempt to rank things (the best movies, the best singers, the best athletes, the best novels, etc.)'.

You are likely to see maximisers constantly changing television channels because there might be something better to watch on the other side, or spending hours in shops as they agonise over the best product to buy. They have difficulties writing emails and texts to people because the tone has to be exactly right. And their thoughts will often be dominated by regret and self-blame because they perceive that, despite their efforts, they have missed out on the best.

I remember trying to choose a holiday in Andalucia with a friend. Once I thought we had settled on a choice, we suddenly reflected on its disadvantages: perhaps it was too expensive and involved spending too much time touring in a minibus. So we looked at less expensive choices. But these didn't have the allure of the first choice, where you stayed in a picturesque villa in the hills. You stayed in multi-storey concrete hotels in busy towns. Perhaps the first choice was best after all. But then ... Life is probably much happier for 'satisficers' (a term coined by the American economist and Nobel laureate Herbert Simon). They are willing to accept things that are simply good enough, without them necessarily being the best, so they don't have this obsession for ranks.

Decision overload

The agony endured by maximisers is partly caused by the difficulty of comparing the choices on offer. When we are choosing a car, one model might be spacious, stylish, full of gadgets and reliable, but it might also consume lots of petrol and produce a bumpy and noisy ride, even on immaculate road surfaces. Another might be quiet, economical and reliable, but it might look rather staid and feel a little cramped compared with the first. Faced with a list of cars on offer, all with different pros and cons like these, how do you rank them and how do you identify the best one?

Psychologists have found that, in situations like this, we struggle to process all the information involved. In particular, we face the challenge of making trade-offs in our heads. How much more legroom would compensate for a car that does ten miles less to the gallon? Would extra gadgets be sufficient to make up for a bumpier ride? To avoid a headache, we resort to simplifications. One strategy people use is to rank the contenders on the one criterion we consider to be most important – reliability perhaps – and forget the rest.[5] If two contenders tie on this criterion, rank them on the second most important – fuel economy say – and so on. This simple method has been given the incongruously lengthy title of 'lexicographic ranking', because it reflects the way in which words are ordered in a dictionary. The worry is that it might lead to a car that's highly reliable, but awful in every other respect.

Another strategy is to set some reasonable limits for each criterion and eliminate all the cars from our list that fail to meet them.[6] So we might get rid of the cars that do less than 45 miles per gallon, then those that have less than four feet of legroom, then those with little space for luggage, and so on, until hopefully there's just one car left. The trouble is that you might be rejecting a car that does 44 miles to the gallon, but is brilliant in every other way.

Some employers use this method, called 'elimination by aspects', when they need to whittle down a large pile of application forms from job candidates to a manageable shortlist. One wonders how many excellent people have never made shortlists because their examination grades were marginally short of some arbitrary standard or because their experience in the relevant line of work was a month short of the number of years hastily determined by a harassed manager.

In some cases we can reduce our mental effort to a minimum by choosing a car whose name we recognise rather than any of those we've never heard of – psychologists call this the recognition heuristic.[7] This can sometimes make sense. Products with recognisable names have often been around for a long time and so have been tried and tested. But it won't work if we recognise all the brands available or if the obscure brand is much better than its competitors, perhaps because it is new and innovative.

In the end it's tempting to resort to gut feeling. Choose the car that will impress our neighbours as we park up against the curb or the job candidate who emitted the right vibes and seemed to be the most convivial. After all, we can always retrospectively conjure up a rationale for our choice so that it looks considered and sound.

Panel paradoxes

Most of the 'of the year' awards for the best book, car, player or employee are made by panels of judges, rather than individuals. For example, the Man Booker Prize for Fiction had five panellists in 2018, while the UK Car of the Year had 27. It's often said that two heads are better than one, so presumably five or 27 heads will be even better and will give us a set of ranks that we can truly believe. All we need to do is ask a panel of experts to rank the contenders, then get each panellist to vote on their favourite, and the winner will be the one that attracts the most votes. Unfortunately, the Marquis de Condorcet, an eighteenth-century French philosopher and mathematician, showed that this can, under some circumstances, lead to a very odd result.

Born in 1743, Condorcet was a man with remarkably progressive views for his time. He favoured female suffrage, condemned slavery, defended human rights, and campaigned against the death penalty. Active in the French Revolution, he eventually fell foul of the authorities, who arrested him after a period in hiding. Two days later he died of mysterious causes in prison. But his legacy persists. As a believer in democracy, Condorcet was one of the first people to apply mathematics to the analysis of voting systems and in 1785 he wrote an essay that demonstrated what is now known as Condorcet's paradox.

Suppose a panel of three judges – Pike, Quinlan and Rogers – have to choose the Gibberish of the Year award. The award is to be presented to the celebrity who is judged to have made the most unintelligible statement in the last twelve months. (In reality, there is no such award, but the Plain English Campaign has a Foot in Mouth award for the most baffling statement made by a public figure – recent winners include Jacob Rees-Mogg, Russell Brand, Donald Trump, Elon Musk and Mitt Romney.) Three contenders have been shortlisted, but to save them embarrassment we will simply call them A, B and C. The judges rank the celebrities as follows (so, for example, Pike thinks celebrity A is the best candidate, followed by B and then C):

Pike:	A	B	C
Quinlan:	B	C	A
Rogers:	C	A	B

To simplify their decision, the judges agree to compare the candidates in pairs. Celebrity A is preferred to B by two judges, so A gets more votes than B. Celebrity B is preferred to C by two judges, so B gets more votes than C. The decision looks as if it's settled – A beats B, and B beats C. But just as we prepare to announce that A is the winner, someone points out that if we compare A and C, then C gets more votes. Condorcet showed that we can find ourselves in a never-ending circle of preferences, known as intransitivity. A majority of panellists prefer A to B and a majority also prefer B to C, but, oddly, a majority prefer C to A as well. Worse still, a clever panel member can game the voting process to ensure that their least favourite is not chosen. They can do this by misrepresenting their true preferences – otherwise known as tactical voting.

Over 150 years later, just after the Second World War, the American economist Kenneth Arrow extended Condorcet's findings in his impossibility theorem,[8] which earned him the 1972 Nobel Prize in Economics. Arrow proved that, where more than two contenders have to be ranked, no voting system can be guaranteed to meet a set of reasonable conditions. These include avoiding intransitivity, not having a dictator in the group, and ensuring that, if every member of the panel prefers one contender over another, then the votes will reflect this. In the brave new world that people hoped to build after the Second World War, Arrow's proof was seen as gloomy. It showed that the design of a perfect democratic system was a pipe dream. It also showed that, while individuals can have coherent preferences, groups cannot. When some experts on a panel prefer one car, and others an alternative, it does not make sense to say that the group as a whole prefers one car over another. Politicians are often heard to say: 'the electors have told us that they want …' But can that be true? After an election in Wales in 2007, one Assembly Member told the press: 'The electors have told us they don't want a single party with a majority. We are all going to have to address that.'[9] When I voted in that election, I don't remember seeing 'No single party with a majority' on the ballot paper.

If voting can't guarantee a perfect ranking, perhaps we should simply get a group of experts around a table to thrash out who or what should win an 'of the year' award. But groups of people meeting together can also behave in very strange ways. Assertive and talkative individuals can dominate the proceedings so that only their views are reflected in the group's final decision. Worse still, in groups where there is a high degree of conformity and no one wants to rock the boat, members can even lose sight of reality – a phenomenon known as groupthink.[10] In such groups, each member feels motivated to speak in support of the position put forward by the group's leader, even if this is patently unwise, erroneous or reckless. Those who disagree remain silent and begin to doubt their own views. In the end, the group can confidently make a decision that would seem crazy to any outsider.

Nowhere has the disturbing power of conformity in groups been demonstrated more forcefully than in a set of experiments conducted by the US psychologist Solomon Asch. They showed that people were even prepared to disbelieve the evidence of their own eyes when other members of the group suggested answers that were clearly wrong.

The participants in Asch's experiments were asked to judge which pair of lines were the same length when a small set of vertical lines were displayed on cards. It should have been a simple task – the other lines had very different lengths. But all the group members, except the one genuine participant, were stooges in league with Asch and were all told to announce, wrongly, that a differing pair of lines was of equal length. Surprisingly, 75 per cent of participants were prepared to agree with the stooges on at least one of the occasions when they were asked to make a judgement. Afterwards, some said they were worried about making a fool of themselves by going against the rest of the group. Others were convinced that the other members must be correct since they were all in agreement.

If groups can distort people's judgements in ways like these, then we should be wary of any 'of the year' award that is confidently announced by the media. When we want to rank things, perhaps the answer is to dispense with human judges. Instead, we could use objective data, together with a formula that converts it into a precise scientific-looking score. In other words, create a league table.

The League Table game

My niece is hoping to go to university and has to decide which ones to apply to. It's a tough choice. These days, universities are in competition for students, so they have developed slick marketing campaigns with seductive pitches to attract them. 'Where you belong', 'The adventure starts here', 'World changers welcome' and 'Dare to be fearless' are a few of the slogans designed to resonate with ambitious 18-year-olds. In a world of slogans and marketing hype, just how can a person choose the university course that's best for them? My niece is depending heavily on league tables.

Most of us can't resist league tables. School league tables, university league tables, league tables of cities that are pleasant to live in – they all sell newspapers and we rush to open the page to find the current status of our alma mater or city. Being associated with a top-three university or the most appealing city in the world contributes to our sense of well-being and it might even get us a better job or increase the value of our house. League tables also save us from that mentally draining task of making trade-offs between the pros and cons of different contenders, thereby providing us with a quick and easy way of making decisions. They look official, scientific and exact, and they impose order, literally, on a chaotic world. League tables will tell you where your country is placed for happiness, corruption, recycling, fitness, creativity in advertising, child well-being, productivity, and overall Eurovision Song Contest performance in the years since the competition's inception in 1956. They can also help you to choose an estate agent, a bus operator, a hospital or a child-friendly restaurant. It's a pity that some of them have fundamental errors.

Many league tables are at least transparent. They provide details of how they've collected their data and how the table has been compiled. But looking under the bonnet to see the underlying mechanism can be tedious – most of us prefer simply to enjoy the quick fix of seeing our favourite contender near the top of the rankings. And therein lies a host of problems.

Take university league tables. The people compiling them for newspapers will want to achieve several objectives. They'll want the league table to be newsworthy, so it will need to contain some surprises. It will also need to have some significant differences from last year's table and from those of other newspapers, otherwise who will buy *my* paper? At the same time, the league table must be credible. To achieve credibility it must not overdo the surprises. In Britain, if Oxford and Cambridge don't appear at the top of the table, or near to it, then people will think there's something wrong with the methodology. Perhaps that explains why, in one league table published in the early 2000s, Oxford's Said Business School came second, despite the fact that the school had not supplied half the data requested by the league table compilers.

In the interests of respectability the compilers will need to collect some 'objective' data on the quality of the university, its departments, or a specific type of course it runs. For a postgraduate business course like an MBA, this might be the average salary students achieve as a result of taking the course, the research rating of the university department or business school, the percentage of international faculty they employ, the percentage of lecturers with PhDs, and so on. Of course, there is no guarantee that these are the measures that really indicate whether a course is any good.

Next, they will need to combine all the measures to get an overall score for each university. The problem is that these criteria might not be equally important, so they will need to attach weights to the different criteria to reflect this. For example, they might allow 60 per cent of a university's score to come from the average salary of graduates, 15 per cent from the university's research rating, only 5 per cent to come from the percentage of lecturers who have PhDs, and so on. But potential students will differ in the importance they attach to the criteria – some may prefer more international staff, others may regard the university's research rating as highly important. All the compiler can do is make a stab at weights they think will be generally acceptable to their readers. They might conduct a survey to see which criteria potential students consider to be important, but they'll need to average these responses. As a result, the weights are unlikely to reflect an individual's preferences. While the university at the top of the table might be the best for a mythical average person, it won't necessarily be the best – or most comfortable – choice for my niece.

But, even setting aside these worries, there is another more profound problem with the weights used in many university league tables – and, ironically, it means that they do not stand up to academic scrutiny. It all relates to the meaning of the word 'importance'. If a potential student is using the league tables to decide where to study, they might regard the average salary following graduation as the most important factor in their choice. But, surprisingly, this might be of little relevance to their choice between universities. Suppose, for a moment, that the average graduate salary was exactly the same for every university. Then it has absolutely no importance in their decision. Wherever they study they'll expect to receive the same salary, so this factor is actually irrelevant. It should therefore have a weight of zero per cent.

More realistically, if there are differences between the average salaries, but the difference between the best and worst is small – say £200 per year – the average salary has some relevance to my decision, but its importance is tiny. Where I study will now make very little difference to my expected future earnings, so its weight should also be tiny. In other words, the weights should reflect not the importance of a criterion per se, but the importance of the *difference* between the best and worst performances on that criterion. Unfortunately, there is no evidence that most published league tables have weights that do this. Yet, if they don't, small differences on some criteria can have a huge effect on the scores if they have been assigned large weights. A university might be near the top of the table just because its graduates have a slightly higher average salary than other universities, despite a relatively poor performance on all the other criteria.

It can even be shown that, under some circumstances, this fault in the weights can lead to a strange phenomenon called rank reversal. University A might have a higher rank than university B in a league table, but if a new university, C, is subsequently added to the table, B suddenly finds that it has a higher rank than A, even though the relative performances of the universities remain unchanged. A simple demonstration of how this can happen can be found in the notes.[11] If university A is performing better than B, then surely this must still be the case whether C is in the table or not.

These concerns about importance weighting are so easy to miss that for many years they even eluded top decision analysts (people who design methods and software to help decision makers) until someone – no one knows who – discovered that many widely used methods were wrong. A seminal book on decision making published in 1986 by two leading American academics, Detlof von Winterfeldt and Ward Edwards, broadcast the bad news, but also showed how the weights should be assessed.[12] Countless decisions must have been made based on apparently scientific methods that were flawed. Worryingly, there is still commercial decision-aiding software around that encourages people to treat the weights in the same way as most league tables do.

The latest university league tables often merit a special supplement in newspapers. 'Find out inside who the big gainers and losers are' is a typical front-page tempter. On later pages there are likely to be interviews with the high-flying university leaders, telling us how they achieved their success, and with star students testifying what a great place the top university is. 'University X storms up the rankings' is a typical dramatic subheading, conveying images of hordes of gowned academics rushing up a hill like an invading army, scattering mortar boards in their wake. Below there will often be a photograph of a smiling vice-chancellor, who'll doubtless think that they merit a pay rise after such a performance. Overnight, university websites will be revised to tell the world that they came top or in the top five or top ten and a flurry of press releases will follow.

But when the differences between universities' overall scores are small – as they often are – minor random events in the course of a year can cause them to move up and down the rankings quite a long way. This year's graduates might report a slightly lower average salary than last year's. Dr Jones moves to another university to live close to his daughter, so the percentage of staff with PhDs is reduced. While the resulting changes in rankings make each year's new league table newsworthy, it just doesn't make sense to believe that the relative quality of universities can change so much over a twelve-month period. For example, in the *Guardian's* university league tables, Loughborough University rose from 11th to 4th place between 2016 and 2017. A quick glance at the 2018 table would suggest that Bath University, in 5th place, is much better than Lancaster University, in 9th position, yet the former scored 81.9 points and the latter 80.8 – a difference of 1.1 points.

Even apparent longer-term trends in a university's status start to seem doubtful. Between 2012 and 2016, Yale University, the stamping ground of over sixty Nobel laureates and the alma mater of five US presidents, slumped from 7th to 15th in the QS World University Rankings. But do the criteria used to determine the rankings really reflect the quality of a university's teaching and research? The average number of times research papers written by staff have been cited by other researchers depends on the mix of subjects offered by the university – science and engineering papers tend to be cited more often than those in the arts and humanities. And a high percentage of staff who come from overseas is at best a debatable indicator that the university is producing more world-class research.

Sometimes tectonic shifts in ranks can result from strange factors. A few years ago, MBA courses at British universities in an international league table rocketed ahead of many American courses. The reason: the average increase in salary received by students on graduation – measured in pounds in UK universities – was converted into dollars by the table compiler. And at that time the pound was riding high on international markets against the dollar. Overnight, British management education had become superior to its American counterpart. No one thought to check the exchange rate.

City rankings that rankle

Pity the people of Vienna in 2016. They no longer lived in the most desirable city in the world. The city of Freud, Mahler, Schubert and Schrödinger, replete with the glorious buildings of the Habsburg empire, had only scored 97.4 in the *Economist*'s ranking of the liveability of world cities. Robert Doyle, the Lord Mayor of Melbourne, which had pipped Vienna to the title, was tweeting that it was a 'Great day to be a Melburnian'. Melbourne's winning score was 97.5, just 0.1 above Vienna's. At least the disappointed denizens of Vienna didn't have to put up with life in Hamburg, which scored a mere 95.0 and languished in 10th place. Worse still, they could wake up one day and find themselves dwelling in London – down in 53rd place.

What a difference that extra 0.1 made. As the American poet William C. Bryant said: 'Winning isn't everything, but it beats anything in second place.' The headlines around the world were all about Melbourne. Rowing crews were pictured as they lined up on the Yarra River against a background of grassy banks, trees and surging skyscrapers that shouted modernity and prosperity. And Melbourne's affluent beach-lined suburb, Brighton, got in on the act. Its residents could be seen tanning themselves on golden sands, fringed by gaily painted beach huts, palm trees and sprawling mansions (median price: £1.6 million).

But it seemed that all was not well in this urban paradise. A press conference called by the Deputy Lord Mayor to promote Melbourne's 'top of the world' ranking was interrupted by a woman shouting: 'It's disgusting!' and 'Melbourne should be ashamed of itself.'[13] She was protesting about the 74 per cent increase, over two years, in the average number of people who were sleeping rough in the city's central business district, which was then estimated to be well over 200 people.

Then there was a survey that had found that nearly half of Melburnians were frustrated by the city's high cost of living.[14] And, while Melbourne had scored a perfect 100 out of 100 for its education, healthcare and infrastructure, the local paper could not resist contrasting this with the experiences of rush-hour travellers on the city's Punt Road or the thousands of people awaiting elective surgery.[15]

Nowhere is perfect, but a score of 97.5 looks pretty close to perfection. It raises the question of whether you can represent liveability (whatever that may be) as experienced by millions of people by a single number measured to one decimal place. It turns out that the *Economist*'s table isn't primarily designed to represent the experience of ordinary citizens in the places it covers. Instead, its main role is to provide guidance to multinational companies when calculating the relocation packages that should be awarded to employees to allow for the gloom or pleasure of moving to a new city.[16] A score above 80, the *Economist* suggests, should attract no extra allowance, 70 to 80 is worth an extra 5 per cent of salary, while 50 or less should earn mobile global talent an extra 20 per cent. All of the criteria used in the ranking are focused on the needs of expats. Its five main categories, with weights in brackets, are stability (25 per cent), which covers factors such as threat of crime and terrorism, healthcare (20 per cent), culture and environment (25 per cent), education (10 per cent) and infrastructure (20 per cent). There's no reference to cost of living, perhaps because this doesn't concern global talent whose salaries already take this into account.

So the league table is actually intended as a ranking of the attractiveness of cities for people who work for multinational companies and who are relocated there. But that doesn't make for crisp newspaper headlines or stories. As a result the distorted perception emerges that it's a measure of the pleasantness of a city for ordinary folk. They might proudly announce 'I live in the world's most desirable city', not knowing that the table is not designed for them. But are the rankings in the table even meaningful to expats? What about aesthetics, friendliness and a sense of community? And look at those weights. The *Economist* does not report how they were obtained. Who is to say that stability should have a weight of 25 per cent, while infrastructure gets only 20 per cent? A person considering moving to a new city might be much more concerned about the awful road conditions on their daily commute than the remote threat of crime in the exclusive neighbourhood where they would live.

Despite their scientific veneer, the weights have the same problems as those used in university league tables. They are arbitrary and, almost certainly, don't reflect the differences between the best and worst performances on each criterion. Make a small change to those weights and, given the closeness of the scores, the presses could be rolling triumphantly in Vienna, or even Hamburg, celebrating their city as the best in the world. In this case, there would doubtless be serious investigations and crisis meetings in Melbourne to establish where it all went wrong. All because of a tweak in an anonymous league table compiler's weights.

The thief of joy

Theodore Roosevelt once said: 'Comparison is the thief of joy.' That chimes with more recent findings in psychology which reveal that, when we compare ourselves to others, the misery of being worse than them far exceeds the joy of outperforming them.[17] Potentially, we have more to lose than to gain through comparison. People carrying out excellent work in universities near the bottom of the table may become disheartened. And their lowly rank may mean that they attract less dedicated students. A spiral of decline can set in. Whether whole cities can feel miserable or joyful is an open question, but no doubt league tables can have an influence far beyond that merited by their methodologies. So is it worth trying to quantify the quality of life in cities at all – that is, taking the measurements, but avoiding the intercity comparisons?

People are likely to have difficulty in relating to cold statistical measures such as those reflecting 'the toxicity of airborne chemicals from local industries – measured on a scale from 0 to 311,000', as in one index,[18] or Gini coefficients, which measure income inequality. This is especially true when a formula using arbitrary weights is used to combine these into numbers like 97.5 or 97.4. If quality-of-life measures have any role – other than rewarding expats – it is surely to motivate people to make improvements to their environment where this is needed. This calls for so-called hot indicators – individual measures (as opposed to scores that try to combine lots of different measures) that people can relate to because they are both easy to understand and interesting.[19] Better still if they have a touch of the bizarre, as they are more likely to attract attention and headlines.

In 2005 residents of the Australian city of Port Phillip – a local government district that, by coincidence, is part of Melbourne – wanted to create more welcoming and friendly neighbourhoods. Someone suggested that they measure 'smiles per hour'.[20] The idea was taken seriously and volunteers were dispatched to designated streets for fifteen minutes, carrying strictly neutral expressions on their faces, to make eye contact with people, and record how often they smiled. The collected data revealed which neighbourhoods had people smiling most frequently and which had a preponderance of less happy faces. The local mayor, Janet Bolitho, argued that smiling helped people to feel more connected and safer, so they had less fear of crime. Was the initiative a success? According to one report, within eighteen months the percentage of people smiling at you in Port Phillip increased from 8 to 10 per cent. That difference may, of course, fall within the range of measurement error, but anecdotal evidence was generally positive. Moreover, being involved in data collection can motivate people to seek improvements. In Port Phillip the so-called 'smile spies' – the volunteers – themselves apparently became catalysts for greater friendliness.

Lots of other hot measures have been used. Seattle calculated the ratio of vegetarian restaurants to the number of McDonalds, the amount of bird seed sold in garden centres as a proportion of the quantity of pesticide, and a variety of similar measures, to assess whether the city was moving towards greater sustainability. In Hertfordshire, people counted the number of streets that were quiet enough for a conversation, while in Colchester the number of stag beetles was regularly recorded.[21] Of course, more technical measures, so-called cold indicators such as CO_2 emissions, have an important role, enabling scientists to alert us to problems that need tackling. But there seems to be no need to combine these with other measures to get an overall score, which only serves to hide the individual factors that need attention.

So why this obsession for combining different measures into overall scores so that cities, universities, schools, countries, and countless other things, can be ranked? Much of it relates to our preoccupation with status. The distinguished Californian-based neuroscientist Michael Gazzaniga has argued: 'When you get up in the morning, you do not think about triangles and squares and these similes that psychologists have been using for the past 100 years. You think about status. You think about where you are in relation to your peers.'[22] Research by another Californian-based academic, psychologist Cameron Anderson, and two of his PhD students, suggests that the desire for status is a fundamental human motive, something that everyone craves even if they are not aware of it.[23] They even found that a person's perception of their own status can affect their physical and mental health. League tables meet our need to establish the relative status of things we are associated with.

Of course, the competitiveness fostered by league tables can sometimes be an effective motivation for organisations and countries to make continuous improvements. But all this assumes that the league tables are accurate, that they are measuring something meaningful, and that rank actually matters. Often they fail on all counts. Worse still, there's an incentive for people to focus on enhancing their position in the table at the expense of their 'true' objectives. In England, the number of primary-school children excluded from schools and sent to special units doubled between 2011 and 2018 as head teachers sought to remove troublesome pupils who might otherwise depress the school's ranking in league tables – a practice known as 'off-rolling'.[24]

Is an apple better than an orange and an orange better than a pear? If you need to know, I could construct a league table that would tell you.

2

Perilous Proxies

Narcissism, boredom and physical attraction by numbers

Are top managers of companies narcissists? And, if they are, does it have any effect on how their companies perform? These questions were posed by two researchers at Pennsylvania State University, Arijit Chatterjee and Donald Hambrick.[1] To the layperson, narcissism is usually viewed as a clinical disorder where a person has excessive self-regard. After all, in some versions of the Greek myth, Narcissus withered away because he couldn't stop admiring his own beautiful reflection in a pool. Today psychologists regard narcissism as a characteristic that can occur in different degrees along a spectrum. It can involve traits such as exploitativeness, a sense of entitlement, a desire always to be at the centre of attention, arrogance and a superiority complex.[2] Chatterjee and Hambrick hypothesised that the more narcissistic the CEO, the greater will be the volatility in their company's performance. Managers with this characteristic, they argued, usually favour bold actions that attract attention, but which can also result in spectacular wins or disastrous losses.

But how do you measure the narcissism of a top manager on a numeric scale? Few such managers would agree to give up their time to take a psychometric test when they might be told at the end: 'You are an extreme narcissist.' And asking employees about their boss would be unlikely to lead to reliable assessments, given the politics of corporate culture. Instead, the two researchers relied on measures such as the prominence of the CEO's photograph in annual reports, the CEO's prevalence in press releases, and their use of first-person singular pronouns in interviews. The research, which was confined to the computer software and hardware industries, did find that CEOs who scored highly on these measures were more likely to involve their companies in big mergers and acquisitions and increase the chances that their firm's performance would oscillate between extremes. But, of course, the findings depend crucially on how well the chosen measures were true reflections of managerial narcissism. As we shall see, proxy measures are by no means guaranteed to act as reliable representatives of the 'truth'.

Measuring narcissism by proxy would have been meat and drink to Francis Galton, the Victorian polymath who believed that: 'Nothing in human nature is indeterminate. Anything and everything can be measured.'[3] Galton, a cousin of Charles Darwin, was an obsessive number lover, who seized the chance to quantify at every opportunity. He counted the number of brushstrokes it took to paint his portrait, and while on holiday in the French town of Vichy in 1888, he secretly recorded the time it took passers-by to walk between two locations, seven and a quarter yards apart, on a street. In some of his endeavours he was aided by one of his inventions – counting gloves, which included a hidden panel he could prick with a pin, allowing him covertly to keep a tally on whatever he was observing. The relative beauty of women in different parts of Britain and the propensity of residents of European cities to tell lies were all assessed via Galton's clandestine pin and panel.

But it was Galton's use of proxy measures that are perhaps the most memorable. He reasoned that you could measure the boredom of a meeting by counting the total number of times its participants fidgeted per minute. When he tested his idea he realised that looking at his watch might influence the results, so he also had to count his breaths to estimate how long a minute was (he determined that he took exactly fifteen breaths per minute). Galton conceded a limitation of his method: 'These observations should be confined to persons of middle age. Children are rarely still while elderly philosophers will sometimes remain rigid for minutes.'[4]

Then there was Galton's idea for measuring how attracted one person was to another. He conceived that a gauge could be linked to pressure pads that would be surreptitiously laid under the chair legs of dinner guests. A person attracted to another guest would, he reasoned, tend to lean towards that person and the resulting pressure reading would, unknown to them, reveal the true extent of their feelings (Galton's measure did not appear to take into account whether a person was hard of hearing). The idea of using physical quantities to reflect psychological phenomena was particularly appealing to Galton. As a man with a large head, he believed that skull size was an accurate indicator of intelligence. He also spent much time unsuccessfully attempting to link facial measures to criminality.

In the number-obsessed modern world, proxy measurements are widely used where phenomena are inaccessible. They also enable unobtrusive measurements to be taken where people might misrepresent themselves to give a good impression or change their behaviour if they know they are being monitored. In companies, rates of absenteeism and staff turnover might be used as proxies for the level of job dissatisfaction. Night-time lights photographed from satellites have been used to assess levels of economic activity in different countries.[5] In Britain, the number of children receiving free school meals is used to measure levels of educational disadvantage in different regions.[6] And on social media, people judge their popularity by the numbers of likes and friends that they have. As we shall see, even government policies, and our perceptions of how successful they are, can depend on proxy measures. But when do proxy measures tell us what we want to know – and when can they seriously mislead us?

Problematic proxies

The most obvious danger of a using a proxy measure occurs when it bears no relation to what it is supposed to be representing. Galton's use of skull size to measure intelligence is a classic example, though, to be fair to him, he later explored other methods based on people's reaction times to various sound and light events created by a pendulum-based device. Some later research has found modest correlations between reaction times and other measures of mental acuity.[7] Nevertheless, even this finding illustrates a problem with proxy measures – if we are using them because we can't measure the true phenomenon we are interested in, how do we know whether or not they are valid? We might end up having to assess their validity by seeing if they correlate with other proxy measures, which also can't be validated.

The danger of using proxies that are unrelated to what we actually want to measure is only heightened by our tendency to see correlations where none exist. The problem occurs when we believe that two things are correlated even though there is no evidence that they are. For example, we may believe that vaccines are associated with autism in children or that high-voltage electric power lines cause cancer. We then tend to recall the occasions that confirm our belief, but forget the many occasions when the belief is not supported. The relatively few children who were vaccinated and then found to be autistic come to mind easily, but we don't register those who were diagnosed autistic and were not vaccinated or the thousands who were vaccinated and were not diagnosed as autistic. Similarly, we may know someone who lives close to a power line and who contracted cancer and discount the wider evidence that the two things are unrelated. If immigration numbers in a country are used as a proxy for the risk of terrorism[8] – as has been the case with some US politicians – or if the remuneration of chief executives of corporations is used as a proxy for their leadership skills, the effects of illusory correlation can be wide-ranging and damaging. Though, in the latter case all is not lost. One study found that a chief executive's remuneration could provide a reliable proxy for their golf handicap, after other factors had been taken into account.[9]

On some occasions proxy measures might completely mislead us. The number of times an academic's publications are cited by other researchers is often used to measure that elusive concept: the quality of their research. However, a few years ago a research paper reported that a highly regarded advanced forecasting method was not as accurate as much simpler methods. The paper was highly cited, not because of its quality, but because it contained errors. Other researchers felt the need to explain that the paper's conclusion was wrong when they reviewed the literature at the start of their own papers.

Misleading indications like these can also occur when the proxy measure, and the thing it is intended to reflect, both increase together up to a certain point, but, beyond this, they move in opposite directions or simply cease to correlate. Mathematicians refer to this as a non-linear relationship. For example, if we used the time a person spends on a task as a proxy for the quality of their work, we might find that up to a certain point more time on the task is associated with better-quality work, but beyond this time, fatigue and frustration set in, so the quality of their output starts to deteriorate. The acreage available for a species of bird to forage might act as a useful proxy for the local population, as long as the acreage is the factor restricting the size of the population. Beyond a certain acreage, other limiting factors might kick in, such as the availability of a breeding habitat.[10]

Perhaps the most common danger of proxy measurements is that they offer only a partial or crude representation of phenomena – faithfully reflecting the aspect that is most easily quantified and ignoring other important features. Using pass rates achieved on examinations as a measure of the quality of teachers does not take into account the ability of their students or the fact that teaching is not just about getting people though examinations. It is about developing children's skills and understanding of the world and preparing them to be effective citizens. Media reports that more than 20 million Americans suffered from hunger in the 1980s were based on the number of people who were eligible for food stamps but did not receive them (in the USA food stamps are provided through a federal programme to help low-income households to put food on the table). However, later research found that many of those not availing themselves of the stamps were farmers who, although their per capita incomes were low, had substantial assets – typically some had farms worth over half a million dollars – and many could live off their own produce. This, and the fact that they were not using the stamps, suggests that they were unlikely to be suffering from malnutrition.[11]

Perils of proxies

In 2009 a scandal erupted around the hospital serving the English Midlands town of Stafford, a place that rarely hits the headlines (its primary claim to fame had been as the birthplace, around 1593, of Isaac Walton, author of *The Compleat Angler*). But now horrific stories were circulating. Desperately thirsty patients had resorted to drinking from flower vases. Several suffered serious falls, unobserved by those who were supposed to be caring for them. It was common for patients to be left humiliated and sobbing by the staff, while relatives felt obliged to take filthy bedsheets home to wash, remove used bandages and dressings from public areas, and disinfect toilets themselves because they were wary of catching infections. There were estimates in the media that between 400 and 1200 patients died between 2005 and 2009 as a direct result of the poor care.[12]

Yet this was a modern airy hospital with pleasant views of green lawns and fields. Opened in 1983, it had achieved elite status as a 'foundation hospital' and had met its financial and waiting-time targets. This, it turned out, was at the heart of the problem. According to Professor Norman Williams, former President of the Royal College of Surgeons, the managers and staff were so focused on hitting targets that they 'forgot why they were there'.

Numeric targets can have a useful role in motivating people and improving efficiency, and in some cases in the British health system they may have worked.[13] But, as the shocking story of Stafford demonstrated, there is always the danger that problems, and even tragedies, can occur when targets are regarded as a perfect measure of what an organisation is trying to achieve – when the proxy measure becomes the objective.

Goodhart's law, named after the British economist Charles Goodhart, states: 'When a measure becomes a target, it ceases to be a good measure.' Turning a proxy measure into a target often reduces the chances that the 'true' objective will be attained, because people will game the measure. At Stafford hospital, targeting proxy measures of the hospital's efficiency and quality of care led to appalling levels of care as managers sought to cut costs and reduce patient waiting times. If the quality of an academic's research is measured by the number of papers they publish, they might be motivated to concentrate on quantity rather than quality – writing lots of mediocre papers rather than a few outstanding pieces of work. The importance of the quality objective is downgraded by the target measure, which therefore ceases to be a measure of research quality. Similarly, when car manufacturers have to report the fuel economy and emissions of their vehicles measured in laboratory tests, they may be tempted to game these tests. For example, they can fit low rolling resistance tyres specifically for the test, remove optional equipment to reduce the vehicle's weight, and detach wing mirrors and tape over external seams to improve aerodynamics. In extreme cases, as with the Volkswagen emissions scandal of 2015, they may even create software that enables the engine to alter its performance when it detects that a laboratory test is in progress. As a result, the test results bear little resemblance to the pollution a vehicle will cause in the streets or on the open road. One study estimated that, for cars produced in 2014, the gap between the laboratory-measured fuel consumption and on-road consumption was in excess of 50 per cent.[14]

Goodhart's law is not the only problem when people are motivated to do well on a proxy measure. If the proxy is used to represent several different objectives, we lose the ability to prioritise these objectives or to make trade-offs between them. For example, a job applicant's average examination marks might be used as a proxy for their likely success if they are appointed. In this case we would choose the applicant with the highest marks. Yet their success at work may also depend on other attributes that may not be related to examination marks, such as creativity, the ability to work in a team, and reliability. By relying on a single proxy measure we forfeit the possibility of balancing a candidate's strengths against their weaknesses. Similarly, if we are managing a hillside conservation area we may wish to achieve biodiversity, protect threatened species, reduce soil erosion, and create visually attractive scenery. It might be tempting to use the area covered by trees as a proxy in determining how well we are achieving these objectives, but doing so would not reflect the needs of species that do not live in woodland. And while fast-growing coniferous trees would provide cover and prevent soil erosion more rapidly, they would not provide the diversity of a slower-growing mixed deciduous forest. By concentrating purely on maximising the area of tree cover we would not have the opportunity to decide on the relative importance of the different things we want to achieve.

GDP: Gross Domestic Proxy?

I did the government a tiny favour a few months ago: I scraped my car against a wall on a narrow cobbled street to let another vehicle past. The scratches and damaged wing mirror cost me a packet to put right. But the repair work will have added to the country's Gross Domestic Product (or GDP).

These days, most governments are obsessed with GDP and the rate at which it's growing. Annual growth of less than 1 per cent suggests a weak economy and a threat to a government's survival; two consecutive quarters of negative growth and you have the misery of a recession. In recent years, China's growth rate has regularly exceeded a dizzying 6 per cent, while in the UK, the Chancellor of the Exchequer took comfort in 2018 that his growth forecasts for 2020 could be revised upwards from 1.3 to 1.4 per cent.[15] However, Britain's GDP did get a £10 billion boost in 2013 when prostitution and illegal drugs were included in the national accounts for the first time.[16] This is because all countries in the European Union have to calculate their GDPs on an equal basis to determine their contributions to the EU budget and some of these activities were not illegal in other EU countries. Nigeria did even better: its GDP in 2014 suddenly increased by 90 per cent when it decided to start counting economic activity in industries like information technology, music, online sales and film production.[17] In an instant, it was Africa's biggest economy. The Greek government had a more creative way of boosting GDP – for several decades they fiddled the figures. In 2006, they aroused suspicion when they announced that GDP was 25 per cent higher than had previously been estimated. Surprisingly, we hear more about GDP than GDP per capita. According to GDP, China's economy is now the second biggest in the world, but its output is shared across nearly 1.4 billion people. Japan's economy has stagnated since the early 1990s, again according to GDP calculations, but its population has been shrinking and GDP per capita reached an all-time peak in 2017.

GDP, once described by the US Department of Commerce as one of the great inventions of the twentieth century, is designed to measure how well national economies are performing. Different countries estimate it in different ways, but typically it involves either adding up all the money that has been spent in a year or all of the money that has been earned. Because of inflation, the figures are adjusted so that output can be measured in 'real terms'. Gathering such huge amounts of data is obviously challenging. In countries like the UK and USA, many sources are used, including surveys of businesses and trade bodies and visits to a sample of retailers by a field force of clipboard-carrying price collectors.

During the Great Depression of the 1930s, governments realised that they needed to measure the size of their economies in order to inform economic policy. Simon Kuznets, an émigré from Belarus, who would later win the Nobel Prize in Economics, is credited with the first attempt to do this. He produced a report for the Roosevelt administration in 1934 that showed that America's economy had halved in size between 1929 and 1932. This gave Roosevelt the evidence he needed to launch his New Deal policies in which the government spent money on major projects to counteract the stagnation in the US economy. But soon there was controversy about what GDP should be measuring – should it simply be a measure of output, which would include production of armaments or other items which bring a 'disservice to society', or should it be a measure of the extent to which an economy contributes to human welfare?[18]

Modern calculations of GDP, in countries like the UK and USA, take the former approach. As can be seen from the inclusion of illegal drugs and prostitution, they offer no judgements about what is good or bad. The outputs of industries that pollute are valued on an equal basis to those of non-polluters. An electric car costing £25,000 contributes the same amount to GDP as a nitrous oxide-emitting diesel car with the same price tag. Indeed, pollution can increase GDP if products are then manufactured to combat it. Supermarkets sell food products that expand people's waistlines in one aisle and products to help them lose weight in a neighbouring aisle, so GDP, and the supermarkets, gain twice.

But in the world of the twenty-first century, it is becoming evident that, even as an amoral measure of output, GDP is becoming a less and less accurate proxy. These days, people regularly take advantage of free digital services such as Google, Facebook and Wikipedia. They use these to plan their holidays, download recipes for meals, communicate with friends around the world and to acquire knowledge. Years ago these activities would have cost money, but now, because they are free, they do not register with GDP. We are generally better off because we have access to these services, but GDP ignores them.

A huge improvement in the quality of electronic products has accompanied the provision of these services. When I bought my first PC in the early 1990s, a friend told me that I would never fill the computer's 40 megabyte hard disk, which seemed massive. Now, for roughly half the same price in real terms, I can buy a PC with a 1 terabyte hard disk – that's 25,000 times more memory capacity than my old PC – and many other vastly improved features. This means that £1000 spent on a PC now brings far more benefits to a consumer than the equivalent amounts would have done twenty years ago, even after inflation is taken into account. Statisticians who produce GDP figures try to take this into account by using what is referred to as hedonic pricing, from the Greek word *hedone*, meaning 'pleasure'. But measuring how much extra quality you get these days compared with the past for each £1 spent is difficult.

Of course, in some cases, quality may have declined over the years. If you're struggling to breathe while standing on a hot overcrowded train, it probably won't make you feel better to know that doubling the number of passengers in a carriage will double the contribution the train journey is making to GDP. And, when statisticians can't measure the quality of outcomes of a service that's being provided, they'll resort to measuring the cost of the inputs. But costly inputs don't necessarily guarantee quality outputs. US spending per capita on healthcare is nearly double the average of other high-income countries, such as the UK, France, Germany and Australia, boosting its GDP. Yet life expectancy is lower in America and its infant mortality rate is higher.[19]

Another feature of the twenty-first-century world in many countries is an explosion in consumer choice. Whereas in the UK we had only three main TV channels prior to 1982, now we can access up to 600, alongside streaming services like Netflix, Amazon and YouTube. There were 160 types of breakfast cereal available in the USA in 1970; by 2012 there were 4945. Over the same period, the number of automobile models increased from 140 to 684.[20] Contrast that with Henry Ford's famous promise, allegedly made in 1909 when selling his Model T car: 'Any customer can have a car painted any color that he wants so long as it is black.' (Ford himself wrote that he made the remark at a meeting, but there is some dispute as to whether he actually said it.) Then there's customisation – people can now have individual products tailored to their particular needs. Furniture companies offer a large number of different combinations of fabrics, legs and section pieces. Notebook computers are available with different random access memory capacities, processors, displays, hard drives and outer finishes. Some economists argue that this increased choice and variety, which GDP ignores, is itself a benefit to consumers.[21]

Overall, GDP is better suited to the industrial world in which it was conceived – a world of mass production and homogeneous products where you could simply add up the number of units emanating from factories and workshops. It does not work well for service industries, which now dominate advanced economies, and the arts. Two Shakespeare plays acted at double speed would, in theory, be worth twice one presented at normal speed, while four hurriedly carved Henry Moore sculptures of mediocre quality would make four times the contribution to GDP of one masterpiece.[22] Nor does it act as a good proxy for economic well-being in a world confronted by an environmental crisis. Exploiting scarce resources today to satisfy the ephemeral demands of fashion or stuff our houses with things we never use might add to this year's GDP, but those resources will be lost to future generations, to the detriment of their well-being. Recently, NASA's satellite photography detected that the Aral Sea in Uzbekistan – once the world's fourth-largest lake – had completely dried up.[23] A prime cause was the water-intensive demand placed on the sea by cotton growers, who mainly export their production: the cotton in one T-shirt can require up to 2700 litres of water. UK households alone discarded 300,000 tonnes of clothing in 2016, much of it having been worn on only one or two occasions.[24] Yet Uzbekistan's – and Britain's – GDP will have benefited from this disastrous trade.

So what's the verdict on GDP? Despite the brilliant early work of Simon Kuznets and all the elaborations that have followed, the measure is clearly far from being a perfect proxy for the economic well-being of a modern country, and this can pose perils. As the former French president Nicolas Sarkozy argued, when people see a gulf between what official statistics are saying and their own experience of life, this can be dangerous because they believe they are being deceived. And, he warned, nothing is more destructive of democracy. Sarkozy was writing in the foreword to a report he had commissioned to look at how GDP measurement could be improved or replaced.[25] It concluded that we were 'mis-measuring our lives'.

All of this suggests three things. First, GDP calculations urgently need a major overhaul to catch up with the twenty-first century. Second, we should view the well-being of the economy through the perspectives of a range of different measures, not just a single number. Finally, politicians and the media should wean themselves off the belief that GDP, or any other measure, can be an exact scientific reflection of how well an economy is performing. We need to flag up its limitations and remember these whenever we hear the discomforting news that a growth forecast has been downgraded by 0.1 per cent.

A barometer of crime

In the Autumn of 2018 newspapers and television news bulletins carried the alarming headlines that crime on Britain's railways had rocketed by 17 per cent in just one year.[26] This sudden rise seemed even worse because it had followed a long trend of steady decreases in crime, both on the railways and elsewhere. You were left wondering why hoards of criminals had suddenly turned their attention to trains as an attractive place to carry out their egregious activities. It's surprising how many theories can buzz through your mind in a few minutes. Perhaps there were fewer railway staff on trains to look after passengers. Perhaps ticket barriers were out-of-order more often these days, so thieves could more easily board trains and, later on, effortlessly escape with their haul. Or perhaps normally law-abiding citizens were being driven to crime by overcrowding and frequent delays and cancellations of trains.

However, while these headlines may have reflected a good day's work for newshounds, a closer look at the figures suggested a less frightening story. Of course, no one should play down the impact of being a victim of crime, and one extra crime is one crime too many. But, as senior police officers were quick to point out, the chances of being a victim of any crime on a train, according to the statistics, were incredibly low: only 19 crimes were recorded per million journeys in 2017–18.[27] The 17 per cent increase meant that the number of recorded crimes had increased by less than three for every million journeys taken by passengers. Moreover, a decade earlier, around 30 crimes were being committed per million journeys – so, since then, the chances of a crime occurring on a typical journey had actually decreased by nearly 37 per cent.

Then there is the issue of how a crime becomes a statistic – it has to be reported to the police and then recorded. The #MeToo movement may have encouraged victims of sexual crimes on trains to be more willing to report offenders. To assist this, and the reporting of other crimes, the police responsible for the railways had introduced a confidential text service for passengers. They suspected that many crimes they would not have heard about in the past were now being reported.

Even after making these allowances, these reported statistics still had one obvious limitation. They ignored the fact that some crimes are worse than others. A murder is clearly more harmful than the theft of a loaf of bread from a shop, so to add these together to make two crimes masks their different levels of seriousness. Statistics that record the harm inflicted by crime would give us all a better idea of the problem than simple numbers of crimes, and they would help cash-strapped police forces to target their resources on the ones that create the most misery. But how can we measure the harm that a crime causes? After all, some victims will be more resilient to the distress caused by crime than others. A burglary accompanied by threats of violence to a house's occupants is likely to be more traumatic than one carried out while its victims are away on holiday. And how do you compare emotional harm to the damage caused to the environment by crimes like fly-tipping or trading in endangered species?

Criminologists have grappled with this problem for many years. One approach is to survey the public, asking them to rate the severity of crimes based on narrative accounts. An American survey conducted in the mid-1980s found that, when this method was used, major crimes were rated as being 300 times more serious than minor crimes, yet its results have never been used by police forces in making operational decisions.[28] Other researchers have used surveys of victims' judgements, but when serious crimes are rare, too few people may be involved to allow the findings to be generalised with confidence.

The latest ideas involve using a proxy measure for harm – the lengths of prison sentences (or number of days of community service) associated with different crimes. Where a fine is levied, this is made commensurate with a period of incarceration by calculating how many days a person on the minimum wage would have to work to pay the fine.[29] The Cambridge Crime Harm Index, developed mainly by Lawrence W. Sherman, an American criminologist now based at the University of Cambridge, uses the starting-point sentence that a first offender would receive if there were no mitigating or aggravating circumstances associated with the crime. These sentences are listed in guidelines issued by the Sentencing Council for England and Wales. This measure has the advantage that it is simple and cheap to use, transparent, and, arguably, reflects society's current view of how heinous different crimes are. For example, assaulting and injuring a victim carries a starting-point sentence of 19 days in jail (though the mean sentence is 184 days).[30] Preparation of a terrorist act where multiple deaths were risked, but plans were only at an early stage, has a starting sentence of 15 years in custody.[31]

The total amount of harm caused by crime can be estimated by adding up the total length of the sentences that would have been levied by courts if the starting-point sentences of crimes that were prosecuted had been applied, rather than the actual sentences given.[32] When the index was used to measure UK crime, it revealed that, while the number of crimes in the decade up to 2013 fell by 37 per cent, the total harm resulting from crime fell by only 21 per cent.

But why take the *starting-point* sentence of an offender as a proxy for the harm that their crime has caused? One reason is that actual sentences reflect factors that are not related to harm. For example, in England and Wales, offenders who plead guilty at 'the first reasonable opportunity' receive a one-third reduction in their sentence.[33] In contrast, repeat offenders will usually receive longer sentences – one study of convicted shoplifters in middle America in the 1980s found that two-thirds of them repeated their crimes at least once a week.[34] Yet the harm caused by a person shoplifting for the first time will, typically, be just as serious as the twentieth offence by a seasoned thief. In addition, some long sentences are imposed to protect the public from potentially dangerous criminals, or to act as a deterrent, rather than to reflect the harm emanating from specific crimes. There is also the concern that some judges might deliver longer sentences because they feel pressurised by ephemeral concerns about particular crimes in the news media.[35] One Israeli study even found that judges tended to be more lenient after they had had lunch.[36]

Nevertheless, actual sentences do also reflect an assessment of the amount of harm a crime has caused – for example, where an assault has caused long-term physical and psychological damage to a victim. The developers of the Cambridge Crime Harm Index justify the omission of the circumstances of particular offences on the grounds that it would be prohibitively expensive to collect sentencing data on each individual crime. But there is an alternative. Britain's Office for National Statistics has developed a Crime Severity Score which uses the *mean* sentence passed on people convicted for each type of crime, rather than the starting-point sentence. An analysis by Matt Ashby, a lecturer in criminology at Nottingham

Trent University, and himself a former police officer, showed that using this approach produced a substantially different picture of the levels of harm caused by crime in England and Wales.[37] For example, domestic burglary caused only 2 per cent of the total harm resulting from crime in 2015–16 in the Cambridge index, but 16 per cent of all harm when the Crime Severity Score was used. In contrast, while rape led to over 36 per cent of total harm in the Cambridge index, it accounted for 20 per cent in the severity score. Differences like these are likely to lead to quandaries for police forces that have to make decisions on where to target their limited resources. They can pose similar dilemmas for governments who have to decide how to share out central funds between police forces in different regions. If they wanted to allocate the greatest funding to regions where crime caused the most harm, then the two methods can give very different indications. For example, Ashby showed that in 2015–16 the West Midlands region ranked third in a table of total harm per citizen on the Cambridge index, but only 17th according to the severity score. In another region, Cleveland, one index estimated that the total harm per citizen was 50 per cent higher than it was in the other.

Attempts to measure the harm caused by crime probably give a better picture of the impact of crime than bald statistics based on aggregate numbers of crimes. But they also show some of the limitations of proxy measures. We can never know the true total amount of harm caused by criminals. Even if every victim gave us an accurate and detailed assessment of the harm they have experienced, can we really quantify and add together an assault victim's mental trauma, a fraud victim's financial loss, and a community's unhappiness caused by littering at a local beauty spot? The reliability of the proxy measures can therefore never be tested against reality. And the discrepancies between the Cambridge and the Crime Severity indexes reveal that trying to view this illusive reality from different perspectives can lead us to see the state of the world in very different ways.

A last hoot at proxies

Half a century ago, before the advent of university ethics committees, a misanthrope could find an ideal outlet by getting a job as a psychology researcher. They could then wreak their vengeance on society, while claiming their mission was solely to enhance scientific knowledge. There was an experiment where students asked people on the New York subway to vacate their seats so they could sit down on trains when there were plenty of other available seats (49 per cent of people complied). Then there's the case of the (fake) theft of beer staged in a liquor store while the owner was out of sight – customers were less likely to report the theft when other customers had witnessed it. Other studies involved persuading people to administer strong electric shocks to slow learners (the shocks were not real and the learners were actors), or, in the famous Stanford Prison Experiment, to behave as brutal prison guards. Some of these experiments had the flavour of the contemporary reality TV show *Candid Camera*, where pranks were played on unsuspecting members of the public. Typical of this genre was a 1968 study designed to create frustration among motorists at road intersections in Palo Alto and Menlo Park in California.[38]

The experiment, conducted by Anthony Doob and Alan Gross of the universities of Toronto and Wisconsin respectively, involved the experimenters sitting at the wheel of a car and deliberately delaying their exit from a junction when the traffic lights turned green. The objective was to establish whether the driver behind was less likely to blow their horn aggressively if the obstructing vehicle was a 'high-status' car rather than an old banger. The posh car was a 1966 Chrysler Crown Imperial hardtop, which had been washed and polished for the occasion. Its driver was smartly dressed in a plaid sports jacket and white shirt. Two 'low-status' cars were used – a rusty 1954 Ford station wagon and an 'unobtrusive grey' 1961 Rambler sedan. Their drivers wore old khaki jackets.

In two cases involving one of the low-status cars, the driver behind hit the back bumper, so the experimenter drove off quickly without waiting to see if the bump would be followed by a honk.

However, in the remaining encounters, 84 per cent of the motorists honked when they were obstructed by a low-status car, but only 50 per cent honked when they were held up by the Chrysler. The researchers hypothesised that the frustrated motorists behind the Chrysler were more inhibited in their response because they inferred that a high-status car meant that a high-status person was causing the delay. Such people, they surmised, typically have the power to impose unpleasant consequences on those displaying aggression towards them. Even though they would be unlikely to exercise these powers in this situation, the following motorists may have recalled other encounters where 'aggression against superiors has been punished'.

The interesting finding was that the drivers in the following car apparently used the model and condition of the experimenters' car, and the way its driver was dressed, as a proxy for their social status and hence their power to exact retribution. Of course, in this case, the nature of the car, and the sartorial elegance of the person at the wheel, had no association with the status of the driver. But we tend to use proxies automatically and unconsciously. Evolution has taught us to look for proxies that allow us to make predictions. A symmetric face in a prospective mate suggests healthy genes; a beard might suggest wisdom; yellow and black stripes in insects indicate danger. Some of these associations yield accurate predictions. Others are myths reinforced by our selective memories – we remember the odd times when the myth was confirmed, forgetting the many occasions when it proved to be false. An expensive-looking car probably is a reliable indicator of the driver's social status, or at least of their wealth, but it is, of course, by no means perfect, and on the occasions when it is wrong (where the driver is an academic psychologist, for example), it can be very wrong.

Sometimes we can't do without proxies. We need to have measurements, but we can't obtain these directly. As Douglas W. Hubbard, the author of *How to Measure Anything*, argues, measurement reduces the uncertainty we have about the world.[39] The Scottish physicist and engineer Lord Kelvin went further, claiming that we cannot understand something until it's measured. But we need to remind ourselves time and time again that proxy measures are not measuring the real in its entirety, and sometimes not at all.

Some organisations hope that we won't remind ourselves. Many of us want to reduce our impact on the planet by choosing environmentally friendly products and we are prepared to pay a premium to obtain them. But we need to beware. 'Greenwashing' – false or misleading claims by a company that its products are good for the environment – is widespread. In a 2007 American survey, conducted by TerraChoice Environmental Marketing Inc., all but one of 1018 consumer products surveyed came with claims that were demonstrably false or designed to mislead.[40] Greenwashing comes in many forms, ranging from crude falsehoods – one brand of dishwasher detergent came in plastic packaging, but the manufacturer claimed it was 100 per cent recycled paper – to promotion campaigns featuring forests, wildlife, mountains and unpolluted lakes. An advertisement for the Shell oil company even had images of flowers emanating from the chimneys of an oil refinery rather than smoke.

In the American study, however, the most common ruse was 'the sin of the hidden trade-off', where one or two green attributes of a product are highlighted in the hope that the consumer will regard these as a proxy for its overall environmental benefits. For example, a major international company claims that its bathroom cleaner, which is sold in enticing green plastic containers adorned with pictures of flowers, is made of 98 per cent natural ingredients, such as citric acid. In fact, the product mostly contains water – admittedly a natural ingredient – but when this was removed by analysts in Canada, a quarter of the remaining contents turned out to be petroleum-based chemicals.[41] Another company emphasises that using its tampons without applicators saves around one pound of waste going to landfill per woman per year. It doesn't advertise that tons of fertilisers, insecticides, herbicides and fungicides are used to grow the cotton that its products contain. And a brand of paper towel is marketed as containing 80 per cent recycled material, but the extent to which its production uses water and energy and causes pollution is not disclosed. The benefits are stated while the costs are concealed, and we consumers salve our consciences, trusting that we have been told enough to judge.

Benjamin Franklin said that half the truth is often a great lie, while Alfred Lord Tennyson said that a lie that is a half-truth is the darkest of all lies. When we rely unquestioningly on proxies, we may be tricked into damaging the very things that we value.

3

One Number Says It All

A questionable quotient?

In October 2017, as the relationship between President Trump and his Secretary of State, Rex Tillerson, descended into an exchange of insults, Trump responded to reports that Tillerson had called him a moron by suggesting that they should both take IQ tests.

'And I can tell you who is going to win,' he predicted with characteristic modesty.

Britain's own blond-haired maverick, Boris Johnson, seems to have a similar attachment to intelligence tests. Speaking about the virtues of inequality, in his role as Mayor of London in 2013, he argued: 'Whatever you may think of the value of IQ tests it is surely relevant to a conversation about equality that as many as 16 per cent of our species have an IQ below 85 ... Over 16 per cent anyone? Put up your hands.'[1]

However intelligent we are, there are limits to our ability to absorb and process masses of digits. Life becomes much easier if we can reduce everything to a single number, and this applies all the more when we encounter phenomena with multiple dimensions, such as intelligence. Why not tidy away all that bothersome complexity? Why not grapple it into something neat and easy to use? The history of IQ testing shows why not.[2]

It began with the best of intentions. The first IQ tests were devised by Alfred Binet, a French psychologist who had abandoned a career in law. In the early 1900s, he was commissioned by the French government to design a method for identifying children with learning difficulties who would therefore require special assistance. Focusing on memory, attention and problem-solving skills, Binet, assisted by Theodore Simon, designed what became known as the Binet–Simon scale. Well over a century later, it still forms the basis of modern intelligence testing. Yet Binet cautioned against using his test as a measure of intelligence, arguing that intelligence was a complex, multifaceted and fluid attribute that could not be captured in a single number.

However, Binet's reservations were cast aside, and his invention, refined, developed and misused by others, soon developed a life of its own. Like an escaped monster in a horror movie, it was soon wreaking a terrible toll on many of those who encountered it. And one of its victims was Carrie Buck.

Born into poverty in Charlottesville, Virginia, in 1906, Carrie was sent to live with foster parents after her mother was confined to the Virginia State Colony for Epileptics and the Feebleminded. Reports say that she was a child of average ability at school, but when she was 17 she was raped by the foster parents' nephew and became pregnant. The foster parents accused Carrie of being promiscuous and out of control. Once she had given birth, they had her committed to the same infamous institution as her mother.

Carrie's misfortune was compounded by the fact that she was living at a time when the eugenics movement was becoming influential. The eugenicists believed that traits such as imbecility and feeblemindedness, as they were then called, were inherited and that, to improve the quality of the human stock, people with these traits should be prevented from breeding. In 1914, twelve US states passed laws permitting the sterilisation of inmates of public institutions.

A prominent advocate of the policy was Harry Hamilton Laughlin, superintendent of the Eugenics Record Office at Cold Spring Harbor, in New York. Because the earlier sterilisation laws had been successfully challenged, in 1924 Laughlin presented a template that could be used as a model for sterilisation laws across the country – laws that could withstand judicial scrutiny and legal objections. Laughlin claimed that, if his law was universally enacted, the genes from 'the most worthless one-tenth of our present population' would be eliminated within two generations.[3] In Virginia, a law was duly passed, but it needed to be tested. Eighteen inmates of the Virginia colony were selected for sterilisation and it was decided to arrange a lawsuit on one of the patient's behalf, challenging the decision, so it could act as a test case. All of those selected were women, and Carrie was one them. It was her case that would go to court in Virginia in the case of Buck versus Bell in 1924.

Those advocating sterilisation needed to establish that Carrie was feebleminded and that the condition was hereditary. To confirm that Carrie and her mother were intellectually deficient, Laughlin used the new Stanford–Binet IQ test and, on that basis, classified them both as 'imbeciles' – a category below 'moron' but above 'idiot'. A Red Cross worker, who had examined Carrie's daughter, described the child as 'abnormal, listless and unresponsive'. Conveniently, for those who believed that feeblemindedness was passed down through the generations, Carrie could now be labelled as a feebleminded daughter of a feebleminded mother and herself the mother of a feebleminded child.

The court found in favour of those requesting sterilisation, and later, in 1927, the decision was confirmed by the US Supreme Court. In its three-page judgement it concluded: 'Instead of waiting to execute degenerate offspring for crime, or to let them starve for their imbecility, society can prevent those who are manifestly unfit from continuing their kind … Three generations of imbeciles are enough.' Carrie's fate was sealed.

In the following years – right up to the mid-1970s – around 60,000 Americans in thirty states were compulsorily sterilised. These included Carrie's sister, Doris, who was told that she needed an appendicitis operation and was unaware of what had happened to her until much later in life. Carrie died in 1983 and is buried near her daughter, who survived only to the age of eight. A plaque in Charlottesville, Virginia, tells Carrie's story and refers to later evidence that she, and many others, had no hereditary defects.

In 1989, the American Association for the Advancement of Science rated IQ tests as one of the twentieth century's most significant scientific discoveries, along with the discovery of DNA, the invention of the transistor, and mechanical flight. By then, billions of IQ tests had been administered in countries around the world. They have been used to determine the most appropriate type of education for children and work for adults. Those recruiting police officers in the town of New London in Connecticut even set a maximum limit for IQ. In 2000, when Robert Jordan, a 49-year-old college graduate tried to sue the police department after being turned down because his IQ of 125 was too high, the courts ruled in the department's favour. The policy was justified on the grounds that high scorers could become bored with the work and quit soon after they'd finished their expensive training. Jordan became a prison guard.[4]

Of course, it's unfair to blame any tool for our misuse of it, and most people today would condemn the horrors of the past. The more relevant question is: does IQ measure what it purports to measure? Were Binet's caveats justified?

The sceptics argue that nobody quite knows what intelligence is and therefore it has never been properly defined. Even in the 1920s, the journalist Walter Lippmann argued that IQ tests were merely a series of stunts. 'We cannot measure intelligence when we have not defined it,' he said. Others argue that the tests focus on particular attributes and so discount people's creativity, intuitive thinking, and their practical and emotional intelligence. It's claimed that they reflect what someone has learned, not their potential, and that they are biased in favour of those brought up in a Western culture. Then there is the process of testing itself. People who are stressed, tired or unfamiliar with the procedure are likely to be at a disadvantage, while those administering the tests, and those interpreting them, can bias results. Perhaps this explains why one's test scores are not fixed – they are likely to vary on different occasions when one takes the test, allegedly by up to 15 points.[5]

So how do the supporters of IQ testing respond? Stuart Ritchie, a research fellow in the psychology department at the University of Edinburgh, argues that there *is* a phenomenon of general intelligence that can be defined and measured accurately.[6] He cites a definition by a fellow psychologist, Linda Gottfredson: 'Intelligence is a very general mental capability that, among other things, involves the ability to reason, plan, solve problems, think abstractly, comprehend complex ideas, learn quickly, and learn from experience ... It reflects a broader and deeper capability for comprehending our surroundings, "catching on", "making sense" of things, or "figuring out" what to do.'[7] Moreover, these different aspects of intelligence tend to be correlated – if you score highly on one of them, you also tend do well on the others. This, says Ritchie, proves that general intelligence does exist.

If intelligence – as measured by IQ tests – has any real meaning and value, it should be predictive of some important aspects of our lives. In his book, *Intelligence: All That Matters*, Ritchie presents an abundance of evidence that it is. People scoring higher on the tests tend to be healthier, live longer, do better in exams and stay in education longer, perform better at work and earn more, have a lower risk of being involved in accidents, and be more socially liberal. Having a higher IQ is also moderately associated with psychological measures of creativity. Of course, many of these benefits will be mutually reinforcing – doing well in exams, for example, will mean a person is more likely to stay in education longer. Nevertheless, most of these correlations still apply after controlling for other factors like a person's social class. But these findings are averages – for example, on average a person with a high IQ will enjoy better health – so they don't necessarily relate to individuals, and there can be quite a sizeable scatter around the averages.

So who is right? Today, most researchers agree that IQ tests have value when they're used for their original purpose, which was to help people, rather than to limit their opportunities. For example, they can be helpful in demonstrating the impact of traffic pollution on children's mental development or ensuring that opportunities are not denied to gifted children from disadvantaged groups.[8] But many researchers would also acknowledge that there are aspects of cognitive ability (a term preferred by some psychologists) that the tests do not capture.

Even for those aspects that are captured, there is now a trend towards dispensing with a single number and replacing it with different scores to represent different cognitive skills.[9] Currently, the most widely accepted view of intelligence is the Cattell–Horn–Carroll theory. This distinguishes between broad aspects of intelligence. Fluid intelligence, for example, is the raw processing power of a person's brain which enables them to solve problems that do not require experience or acquired knowledge. It is therefore relatively free of cultural influences. Crystallised intelligence relates to cognitive abilities acquired through experience, enabling someone to perform well on vocabulary or general knowledge tests, for example. While fluid intelligence peaks around a person's mid-twenties, crystallised intelligence will usually go on increasing until they are well into their seventies and even beyond. Other broad aspects of intelligence include the efficiency of a person's memory and the speed at which they can process information. The Cattell–Horn–Carroll theory goes on to recognise that these broad categories can themselves be subdivided into over seventy narrower abilities such as mathematical ability, listening skills, foreign language aptitude, visual memory, musical discrimination and judgement, and memory span.[10]

As we have seen, the scores on tests can vary over short spaces of time, depending on factors like the level of anxiety and motivation of the person taking the tests and the skill and experience of the person interpreting the results. Ideally, test results should therefore be presented as ranges – reflecting the margin of error of the scores – rather than spuriously accurate exact figures.

All of this suggests that, if someone proudly tells you they have an IQ of 156 as if this number is an immutable, scientifically exact and all-informative representation of their mental prowess, they are not being terribly intelligent.

The lure of averages

Averages offer another convenient way of using just one number to represent masses of others. For example, in 2016 a survey by the American company Reebok based on 9000 people in nine countries concluded that the average human lifespan is 25,915 days, but that the average human will spend only 0.69 per cent of this time exercising, though apparently they will also climb the equivalent of Mount Everest 45 times.[11] In case you need to know, other researchers have found that the mean (average) height of men aged 21 in the Netherlands is 6 feet and half an inch (183.8 centimetres). In neighbouring Belgium, the corresponding figure is two inches less (178.7 centimetres).[12] In Singapore, the average time it takes a pedestrian to walk 60 feet is only 10.55 seconds. In Blantyre, Malawi, people on average complete the distance in a stately 31.60 seconds, according to a study in 2006.[13]

When you want to be deliberately vague while at the same time conveying an impression of mathematical precision, there is little to beat the term 'average'. Ideally, an average is a value that is typical and representative of a set of numbers – but it comes in three common forms: the mean (or arithmetic mean, to be precise), the median and the mode. (For the mathematically inclined there are also trimmed means, Winsorised means, geometric means and harmonic means: means fit for any end.) And though they are all called 'average', these three measures can yield very different values.

Suppose a company has four workers and a relatively highly paid manager and their weekly wages are: £200, £200, £300, £400 and £5000. The mean is the total wage bill divided by the number of people, that is, £6100/5 or £1220. This is clearly not typical of anyone's wage because the mean has been distorted by the extremely high wage of the manager. It's the same as saying that on average you'll be comfortable if you put one foot in a bucket of freezing water and one in a bucket of boiling water. The median is the middle wage when the wages are placed in ascending order. Here it's £300, so in this case, it's more typical at least of the workers' wages. Finally, the mode is the wage which appears in our list most frequently so it's £200. All of these values can be referred to as the 'average wage'. If there's a dispute about the level of wages at the company, this means that each side could quote the average that best suits their purposes. I once taught a course to trade unionists called Statistics for Bargaining. When I inadvertently overran a class by ten minutes I was worried that I might have an industrial dispute on my hands, but as it happened the attendees were so pleased to learn about the multiple meanings of 'average' that they were unaware of the intrusion into their lunch break.

Another problem with averages, however they are measured, is that it's tempting to forget that individuals can vary considerably around an average. Treating everyone as if they are average can lead to some strange results.

In the USA in the early 2000s, psephologists – people who study elections and voting statistics – faced a puzzle. Traditionally, rich people were more inclined to vote for the Republican presidential candidate, while poorer people favoured the Democrat. But, in recent elections, richer states had tended to vote Democrat, so they now appeared as blue on the US election map, while less wealthy states were red, indicating a majority for the Republican. So what was going on? Had there been a fundamental change in American politics causing rich and poor citizens to swap sides? The puzzle was deepened by the fact that, if you looked at individual states in isolation, the richer a voter was, the more likely they were to vote Republican. In Mississippi, Ohio and Connecticut, for example, there was a clear pattern in the 2000 election that the higher a person's income, the greater the probability that they would have cast their vote for George W. Bush, the Republican candidate. And these states were not untypical.

The answer to the puzzle lies in what statisticians call the ecological fallacy, a term coined in 1950 by W.S. Robinson of the

University of California, Los Angeles.[14] The fallacy is committed when we make an assessment about an individual person based on the group to which they belong. Tom lives in Connecticut, where the median income in 2015 was $71,346, so we infer that he is richer than John who lives in Mississippi, where median income $40,593. Effectively, each individual is regarded as being the same as the average for their group, despite the variation around this average. This can lead to the danger of stereotyping – people's judgement of an individual's intelligence, driving skills, personality and political views might all be erroneously based on where they come from, their social class, their gender or their age. According to Walter Lippmann, who invented the term, stereotyping meets our need to simplify our big, complex and ever-changing environment in an attempt to make sense of it. He argued that we actually experienced an 'interior world' populated by stereotypes, as distinct from our real environment, and it was this inner world that determined our actions and views. This was a problem for truly representative government, he believed, because public opinion amounted to nothing more than the collection of popular stereotypes that happened to be circulating within society.

That isn't the only problem it poses for democratic politics. For example, it can lead to the paradox seen in the US elections, where the correlation between wealth and a tendency to vote Republican, seen when people were aggregated at the state level, was completely reversed if they were analysed as individuals.[15] While higher *average* incomes of states was associated with a higher percentage of people voting Democrat, *within each state* more income was associated with a greater tendency of an individual to vote Republican. (A simplified example showing how this can happen is given in the notes at the end of the book.[16]) Many other examples of the ecological fallacy have been identified over the years. In 1897, David Émile Durkheim, a founding father of modern social science, reported in his book *Le Suicide* that Protestants had higher suicide rates than Catholics. He believed this was because Catholics were subject to stronger social control. But Durkheim's conclusion was based on aggregate data – he had found that predominantly Protestant countries had higher suicide rates than those that were mainly Catholic. This did not necessarily mean that an individual Protestant was at greater risk of committing suicide than an individual Catholic.[17] Indeed, the opposite could have been true. Misconceptions like this can be particularly worrying in health research. A study in the 1970s found that, in countries where fat was a larger part of people's diet, women had higher rates of breast cancer.[18] The implications for the danger of a high-fat diet are seductive, but we would be unwise to conclude from this country-level evidence *alone* that individual women are more at risk of the disease if they consume more fat.[19]

Assuming that averages are the rule can also lead to unhelpful psychological interventions and can oversimplify beliefs about how people will react in particular circumstances. Researchers tell us that people are happier when they are married or cohabiting when they are young or old, but not when they are in their forties or fifties, have religious faith, a job, lots of social contact, and are wealthy and healthy.[20] But these conclusions are based on averages and may not apply to any particular individual.

When the psychologists Anthony Mancini of Pace University in New York and George Bonanno of Columbia University examined people's reactions to marriage, childbirth, divorce, bereavement and traumatic injury, they were 'startled' by the variation in their responses.[21] Their representative sample of 16,000 people found that things were far from being the same for everyone. Following divorce, nearly 10 per cent said that their well-being had increased substantially, which makes sense given that happy marriages presumably don't end that way. Then again, 19 per cent of people said they experienced a decline in satisfaction with life, and 72 per cent reported that the end of their marriage didn't change their life satisfaction one way or the other.

Perhaps more surprising was people's reaction to the loss of a spouse. The conventional view is that people typically experience a sharp fall in life satisfaction, followed by a gradual return to the level they enjoyed before the loss. But this pattern was only reported by about one in five of the people in the sample. The majority of those grieving – 59 per cent – showed 'a remarkable degree of resilience'. Their levels of life satisfaction remained stable before and after their partner's death. Mancini and Bonanno worry that interventions which assume that average reactions are the norm can be misplaced. For example, perhaps grief therapy only benefits people who have continued high levels of distress. For others, the majority even, it might actually be harmful.

What to make of all this? While it's less taxing on our brains to see the world through single summarising numbers, we should never forget that they are simplifications. The author Nassim Nicholas Taleb reminds us that a tall person can drown crossing a river that has an average depth of only four feet. You could find yourself freezing on holiday in a resort where the average temperature for that month is 33 degrees Celsius (91 degrees Fahrenheit). A doctor's surgery might be surprised to receive complaints about long waits from patients when its average waiting time is only three minutes. And a talented student might be denied access to a college because they performed badly on an IQ test on a day when they were suffering from hay fever. When we try to capture reality in a single number, we might find that it doesn't represent anyone or anything. Those headline numbers are formulated by throwing a lot of information away – and in many cases, that's the information we need.

4

Leaps and Boundaries

Hooked on 5 per cent

No one thought that Robert Ernst was a candidate for a heart attack. Having given up smoking and drinking in his thirties, the 59-year-old produce manager at Walmart in Texas was regularly running marathons and cycling. He even worked as a personal trainer in a gym, where he met his future wife, Carol. According to Carol, 'the minute the sun was up [Robert] was ready to go for the day.' Her life became one that was full of fun and surprises. The couple camped out for days at balloon fiestas and volunteered to help build houses for the charity Habitat for Humanity.[1]

Robert's only health problem was tendonitis in his hands, so he started to take ibuprofen. Then his doctor prescribed a painkiller called Vioxx. The drug's apparent advantage was that it relieved pain while causing fewer gastrointestinal problems than alternatives. For a time, it appeared to be working and Robert reported no adverse reactions. Then, one afternoon in May 2001, after taking the Vioxx for about six months, Robert told Carol that his pulse seemed to be slower than normal. He had just completed his daily run. That evening the couple dined at the Italian restaurant where they had had their first date, watched some television, and went to bed. During the night, Robert's breathing became abnormally slow. Carol called the emergency services, but they could do nothing. Robert never regained consciousness. Puzzled as to why such a fit, active, fun-loving man should die so suddenly, she became convinced that he had been a victim of a shocking cover-up by the giant pharmaceutical company Merck, which marketed Vioxx.[2]

By 2003 Vioxx was being sold in over eighty countries and was earning Merck $2.5 billion a year. The company saw it as a new blockbuster product and spent $300 million to launch it in 1999. Their sales team received extensive training, which included being taught how to smile and speak when trying to persuade doctors to prescribe the drug. Even Martin Luther King's 'I had a dream' speech was summoned as part of the marketing drive. '[Dr] King was someone with goal-focus,' the training manual declared. 'He kept getting shut down but kept going ... Just as with a physician, you must keep repeating the compelling message and at some point the physician will be "free at last" when he or she prescribes the Merck drug, if that is most appropriate for the patient.'[3]

But not all was well at Merck. During a clinical trial, an unnamed 73-year-old woman, who was taking Vioxx, had died suddenly of a heart attack. In fact, it later turned out that seven other patients had died after taking the drug during the same trial, which was comparing the efficacy and safety of the drug with an alternative called naproxen. Only one person taking naproxen had died during the trial, which involved around 5500 people – split roughly evenly between the two drugs.

If, in a randomised trial, eight people die taking one drug and one dies taking the other, we naturally ask: 'What are the chances of this occurring if the drugs are really equally safe? Is this too much of a coincidence?' In statistics, so-called significance tests are designed to provide guidance on questions like this. Formally, they involve formulating a hypothesis – in this case that the two drugs are equally safe. We would then ask: how probable is it that the death rate for one drug will be at least eight times that of the other *if* the hypothesis is true? The probability in question is called a p-value. Suppose that the probability turned out to be 40 per cent, would you have doubts about the hypothesis that the drugs were equally safe? I expect most people would judge that the difference in death rates could well be due to chance: perhaps when people were randomly assigned to the two drugs, less healthy people just happened to be allocated to the Vioxx group. Toss a coin ten times and you probably wouldn't suspect it of being biased if it gave six heads and four tails. But what if the probability for the trial turning out that way was only 2 per cent? We could then say: if the drugs are equally safe, it is very improbable that we would get eight times more people dying taking one of them. At this point we would start having serious doubts about the 'equal safety' hypothesis and we would probably reject it. The question is: how low does the probability, or *p*-value, have to get before we should reject the hypothesis? Working in the 1930s, the eminent statistician Ronald Aylmer Fisher made a suggestion that it should be 5 per cent, or 1 in 20. Following Fisher's suggestion, when a result has a *p*-value of less than 5 per cent (or 0.05) it is said to be statistically significant. In essence, when the *p*-value falls below 5 per cent, the significance test would be saying that this result is pretty unlikely if your hypothesis or claim is true, so we have good reason to doubt your claim.

It is vital to appreciate that the 5 per cent threshold is arbitrary. There is nothing scientific about it. When asked later why he had recommended that a hypothesis could be rejected if the *p*-value was less than 5 per cent, Fisher conceded that he had no rationale. He had chosen 5 per cent simply because it was 'convenient'. If you obey Fisher's suggestion slavishly, you would reject the hypothesis that a coin is fair if it gave nine heads in 10 throws (it can be shown that the probability of nine or more heads is only 1.07 per cent *if* the coin is fair), but not if it gave eight heads (the probability of eight or more heads *if* the coin is fair is 5.47 per cent). But despite Fisher's suggestion, you are surely entitled to conclude that a hypothesis is untenable if the *p*-value is 6 per cent, or even 10 per cent – it is a matter for your own judgement. This is especially true if 'accepting' an untrue hypothesis is dangerous or costly, as would be the case, for example, if we wrongly accepted that a drug was safe.[4] And yet, despite the absence of a rationale, the 5 per cent threshold has taken on a life of its own. For ninety years it has dominated scientific disciplines such as medicine, pharmacology, psychology, management and education research, and some branches of biology. The phrase '$p < 0.05$' is strewn across papers in scientific journals, indicating that the authors have found a statistically significant difference between two things that allows them to conclude that the difference cannot be explained by chance: Medicine A cures you more quickly than Medicine B; management practice X leads to higher productivity than management practice Y; children learn more effectively if they are taught using the latest method rather than the traditional method. All of this is much more exciting than research that finds there is no evidence that a new medicine is better than the one currently being prescribed, or that there is nothing to suggest a much-hyped new teaching method improves children's reading skills. Research journals are more likely to publish exciting results. One can imagine researchers holding their breath in the short seconds that a computer takes to process their data in the desperate hope that the magical '$p < 0.05$' will appear on the screen. Successful research careers depend on publications and publications can depend on '$p < 0.05$'.

I remember once reviewing a paper for a highly reputable psychology journal. 'Who do you think you are?' wrote the editor in his comments to the hapless authors. 'You claim that you have found significant results when your *p*-values are only just below 10 per cent. The scientific standard is that they should be below 5 per cent.' 'Oh the agony!' groaned one presenter at a research conference I was attending, receiving muffled, but understanding, chuckles from the audience. Her PowerPoint slide was showing a *p*-value of 0.053 (or 5.3 per cent), so she felt that she couldn't claim that she had made a significant finding in her research.

Of course, it's nonsense that a finding should suddenly transition, like an on–off switch, from being scientifically uninteresting when the *p*-value is 5.3 per cent to something that is scientifically 'significant' if the *p*-value is, say, 4.9 per cent, especially as the calculation of *p*-values is usually based on approximations and assumptions that are not strictly true. It's surely better simply to state the exact *p*-value and let the reader judge the plausibility of a hypothesis on that basis. Indeed, in later years Fisher himself recommended this practice, but by then bad habits had been established and the 5 per cent fixation was well ingrained into the practice of many disciplines.

However, the data analysts at Merck had a different reason for being fixated on 5 per cent. Unlike most researchers, they did not want to find a difference between the two things they were examining. They wanted to establish that Vioxx, the newcomer on the block, was just as safe as the widely prescribed option, naproxen. Their problem was that the analysis showed that, if the drugs were equally safe, it was improbable that eight people taking Vioxx in the trial would die of a heart attack, while naproxen would be associated with only one death. In fact, this was so improbable that their *p*-value was well below 5 per cent, suggesting that the hypothesis of equal safety should be rejected. There are reports that this result caused panic among Merck executives. The trial in question had actually been set up by the marketing department to demonstrate that Vioxx led to fewer stomach problems than naproxen. But this had already been established in an earlier trial – the current trial was simply intended as a promotional tool to involve 600 doctors and introduce the drug to them. The reports of deaths were therefore inconvenient, to say the least. Edward M. Scolnick, Merck's senior scientist at the time wrote: 'This course is just stupid. Small marketing studies which are intellectually redundant are extremely dangerous.'[5]

Under pressure, the data analysts pored over their results hoping that the bothersome *p*-value would somehow scrape in above 5 per cent. What if they removed one of the Vioxx-related deaths from the data? Admittedly, the 73-year-old woman had called her son complaining of chest pains, but a heart attack was only the most likely, rather than the certain, cause of her death. It was decided to record her death as 'cause unknown'. Did that get them over the line? Not quite. In fact, the analysts allegedly ended up removing three Vioxx-related heart-attack patients from their data.[6] This meant that only five times more people appeared to die from cardiac problems after taking Vioxx, well within the bounds of chance, as customarily understood. A paper reporting this fortunate finding was published in a respected journal, *Annals of Internal Medicine,* listing Jeffrey R. Lisse, a rheumatologist at the University of Arizona, as the first author. (Lisse later claimed that the paper was actually written by Merck employees, a dubious practice to say the least.[7])

More than four years after Robert's death, Carol took the stand in a court convened to decide on the lawsuit she and Robert's two children had brought against Merck. By now there was growing evidence that Vioxx was responsible for heart attacks. A host of other cases were pending, but Merck vowed to fight every one of them. Its advertising slogan for the drug had been 'for everyday victories', but in this case there was no victory for Merck – at least at first. After a day and a half's deliberation, the jury awarded Carol $24.5 million for mental anguish and economic losses, and also decided that she should receive an additional $229 million as a punishment to Merck for recklessly marketing Vioxx, despite realising its dangers. But Carol's tribulations were far from over. Merck appealed and, in 2008, a three-judge Texas panel reversed the original decision, saying that Carol's lawyers had failed to prove that Vioxx had caused her husband's death. The US Supreme Court refused to change this decision in 2012. Now she was to receive nothing.

Despite this, there was plenty of evidence in other cases that Vioxx increased the risk of heart attacks and Merck eventually relented to pressure and pulled Vioxx from the shelves in September 2004. By this time around 20 million Americans had been taking the drug, alongside millions of other patients in other countries. Assessments of the death toll caused by the drug vary. One medical journal estimated that 88,000 Americans had had heart attacks as a result of it, and 38,000 of them had died.[8] The conservative former publisher Ron Unz noticed that 1999, the year that Vioxx was introduced, saw the largest ever rise in American mortality rates, while the largest fall coincided with its withdrawal in 2004.[9] We should be very careful about inferring causality from correlations (see Chapter 10), but Unz thought that it was possible that Vioxx had killed half a million Americans – nearly ten times the number of American military deaths in the Vietnam war. Finally, in 2007 Merck announced that it would pay compensation totalling near $5 billion to resolve thousands of lawsuits. Although this was probably the largest settlement of its kind, the company refused to admit fault.

Of course, you can't blame Fisher's 5 per cent recommendation entirely for the 'economical' use of data by the Merck analysts. But, by creating an arbitrary dividing line between the conclusions 'the drug is safe' and 'the drug is unsafe', it provided them both with a cover under which they could hide the real probabilities, and an incentive to manipulate their numbers. Even after they had removed the three heart-attack cases, there was still, I estimate, only a 12 per cent probability of there being at least five times more deaths from those taking Vioxx rather than naproxen if the two drugs were equally safe. That probability may have been too low for many people to attribute the difference to chance – after all it's less than one in eight – but the convenient '$p < 0.05$' rule allowed Merck simply to report that there was no statistically significant difference between the death rates of the two drugs – implying, wrongly, that they were equally safe.

This manipulation of data to get the result either above or below 0.05, depending on want you want to show, now has a name: '*p*-hacking'. In a survey of psychology papers, Uri Simonsohn, of the Wharton School, University of Pennsylvania, found that *p*-values tended to be concentrated strangely close to 0.05, suggesting that *p*-hacking is widespread. However, he argued that this does not mean that all scientists are charlatans, deviously manipulating their data into the late evening in darkened empty labs long after their colleagues have gone home.[10] Much *p*-hacking results from the many decisions that have to be made when analysing research results. Should more data be collected? Which method should I use to analyse the data? Are some observations so unusual that they should be excluded? For example, perhaps a respondent accidently clicked the wrong button in an online questionnaire – everyone else gave a score of at least 90 per cent when rating the benefits of a new medical treatment, and this person was the only one to give it a score of zero. If a scientist genuinely believes that a hypothesis is true, they will naturally be biased towards making judgement calls that favour their hypothesis.[11] If I include the possibly confused respondent, I get a *p*-value of 0.14. If I drop them I get 0.03, so I choose to exclude them.

The question remains: why do honest scientists, who are usually intelligent people with enquiring minds, go along unquestioningly with the '$p < 0.05$' rule? One reason is that scientists are often not trained statisticians and many don't understand what a *p*-value is – they just assume that this strange number has to be below 0.05 to make a result worth reporting. In one study, psychology students and their teachers were given a series of six statements that wrongly interpreted the results of significance tests. All of the students and 90 per cent of their teachers believed that at least one of the statements was true.[12]

A second reason is that a simple rule makes for an impactful and easily assimilated story. And, as we've seen, even dour academic journals prefer a good story. 'Method A was significantly better than method B in improving children's spelling' reads better than 'If the methods are equally effective, there is only a 4.2 per cent probability of finding a difference in spelling accuracy at least as great as the one observed here.' That statement also wouldn't get past many newspaper editors when they compile their latest reports on scientific findings. The first statement also has the attraction that it includes the word 'significantly'. Significant implies 'important' or 'worthy of attention'. Having read the statement, we would be likely to presume that method A led to major improvements in children's spelling and, if we have children, we may wonder why it's not being adopted by their teachers. But the major improvement may be an illusion. Statistical significance simply means that the difference between the two methods is unlikely to be due to chance. In reality, the difference between the two methods may be tiny. For example, method A may have led, on a spelling test, to an average of 4.13 mistakes per child and method B to 4.15 mistakes. Differences which are statistically significant, but minuscule, are particularly likely to occur when we have carried out tests on large numbers of people or objects (e.g., if 10,000 children were instructed using each method and then tested). Like a huge microscope, large samples are able to detect minute but statistically significant differences between groups that are of no practical interest at all.[13] We should therefore be interested in the size of the difference, not merely that there is a difference. This is often not reported.

One consequence of all of this is the flip-flop nature of reports we get in the media about what's good or bad for us. One report tells us that drinking coffee will damage your health, another that it will improve it. A glass of wine a day? Avoid it, says one report. Enjoy it and think of the health benefits it brings, says another. Fish, cheese and red meat have all been subject to these conflicting reports.[14] The health-conscious shopper must be driven mad constantly changing their diet as the latest advice hits the headlines.

A third and final reason is that the mechanical nature of the '$p < 0.05$' rule saves scientists the bother of having to think hard about their result – it's significant or not, end of story. Alongside this, the universal application of the rule can give scientific disciplines and their journals the appearance of consistency, rigour and objectivity. Attempts to wean journals off their addiction to the rule have generally met with failure. One of the attempts in the mid-1990s by the British Psychological Society made little progress. 'It just petered out,' according to one of the people involved. 'The view was that it would cause too much upheaval for the journals.'[15] There have been regular attacks on the rule since then, and in 2016 the American Statistical Association felt the need to publish a statement produced by more than two dozen experts warning that its use was 'an especially pernicious statistical practice'.[16] Other experts warned that it was leading to a considerable distortion of the scientific process.[17] One well-known essay written in 2005 by a Stanford professor, John Ioannidis, even had the alarming title: 'Why Most Published Research Findings Are False'.[18]

So will science practice change any time soon? Fourteen years after Ioannidis's article the old habits still hold sway. So I wouldn't bet on it ($p < 0.0001$).

A degree of obfuscation

If your health or understanding of the world through science has not been adversely affected by arbitrary numeric boundaries that mask true underlying scales, then your career might well have been. Getting the career you want often depends on achieving the right exam results either at school, college or university. Usually exams, and other assessments like coursework, are marked on a 0 to 100 per cent scale, but once these marks have been obtained they are often converted to grades or classifications. In British universities, for example, a student's marks will be averaged across the different subjects they have studied and this average will then determine the classification of degree they will be awarded. An average of 70 per cent or more will typically lead to a first-class honours degree (or a 'first'), signifying a star performance; 60 to 69 per cent will lead to an upper second-class honours degree (or a '2:1' on the archaic scale that persists in UK universities); 50 to 59 per cent will earn a lower second-class honours degree (a '2:2'); and 40 to 49 per cent a third. Below that, people will either have deemed to have failed, or be awarded a 'pass degree', though, according to some sources,[19] Oxford University awarded the former British prime minister Alec Douglas-Home a fourth-class degree in 1925.

So established have these classifications become that they even have a set of Cockney rhyming slang phrases to describe them. A first is a 'Geoff', named after Geoff Hurst, the England soccer player who scored a hat-trick in the 1966 World Cup final. A 2:1 is an 'Attila', after Attila the Hun, while a 2:2 is a 'Desmond', after Desmond Tutu, the South African cleric and anti-apartheid campaigner. A third is a 'Douglas', after a former British home secretary, Douglas Hurd, who actually got a first. But a few moments of academic reflection would surely reveal that replacing a student's exact average mark of 61 per cent with an 'Attila' or an 85 per cent average with a 'Geoff' serves only to change exact information into a crude and fuzzy reflection of their performance.

One obvious problem is that there is a tiny difference between a student with an average mark of 69 per cent and one with an average of 70 per cent, yet they will be awarded different degrees, which may have a profound effect on their futures. In contrast, there is a large gap between the 69 percenter and a student scoring an average of 60 per cent, yet they will both be awarded the same degree. Aware of the potential injustice to the student who might just miss out by a hair's breadth on a higher degree classification, many universities have complex formulae that allow some of those on the margin to be upgraded. Of course, even these upgrading rules must be arbitrary. It's unlikely that anyone will have looked carefully at a large sample of past students in the upgrading category and established that they are of the same standard as those automatically getting a higher degree classification.

In some cases, the formulae are open to interpretation or include the helpful word 'normally', which allows decisions to deviate from the norm. As a result, academic staff at examination boards are likely to spend hours debating and agonising whether John Smith's degree should be raised to a Desmond or whether Mary Jones is really worthy of a Geoff. In the end, much may depend on the force of personality of those making the arguments. A board attended by a different set of academics might well have reached a different verdict. The pity is that all this effort and all of those complex rules are entirely unnecessary – all they do is change exact results into poorer-quality information.

In the light of this, many potential employers now insist on access to transcripts detailing students' individual marks on different subjects. But for students wishing to progress to postgraduate study a 2:1, or even a first, is likely to be essential, whatever the individual marks say, so this gratuitous and old-fashioned obfuscation can still hold considerable sway.

In the UK, this habit of converting marks to categories or grades persists at earlier stages of the education system. For GCSEs (General Certificates of Secondary Education), taken mainly by 16-year-olds, though not in Scotland, percentage marks are currently converted to a 9 to 1 scale (9 being best). Before this it was A* to U (A* being best). Even earlier, on the forerunners of GCSEs, it was 1 to 9 (9 being worst, as in my French exam). Besides causing mass confusion among parents and employers, these grades suffer from similar problems to those of degree classifications. Some may argue that, because the toughness of exams varies from year to year, the grades ensure consistency in the reporting of results over time. When the exam is hard, a mark of 65 per cent might merit the top grade, while in easier years, 75 per cent might be needed. In 2017 students were awarded a pass grade, C, on a mathematics GCSE if they got a mark of just 15 per cent.[20] But rescaling the percentages, or adjusting the marking scheme, as long as this is justified, might be a better option. Reporting percentages would be clearer than relying on an arbitrary and confusing numeric grading system.

The UK is far from alone in converting original marks into less informative grades. Wikipedia details the conversion systems in over eighty countries, and in some countries, such as the USA and Canada, there are different systems in different regions or even in different educational institutions. Some systems are complex and the exactness of their boundaries gives the conversion process an aura of scientific precision that is surely illusory. For example, one system used in some US schools awards a grade A+ for marks in the range 96.5 to 100, an A if the mark is between 92.5 and 96.49, an A– for 89.5 to 92.49, and so on. In a mobile world, where it's common for students and graduates to move within, and between, countries, these confusing and inconsistent conversion schemes look like a numeric Tower of Babel.

Lumbered with a label

Arbitrary numeric boundaries can affect the way we see the world more than we might realise. In one of their papers,[21] the psychologists Myron Rothbart, Carene Davis-Stitt and Jonathan Hill recount an old Yiddish tale about a peasant whose land was close to the border of Poland and Russia. Over the years the position of the border had often changed as international conflicts pushed it in one direction or the other. Unsure which country his farm was in, the peasant decided to hire a surveyor to resolve the matter. After several weeks of careful investigation, the surveyor announced that the farm was just inside the Polish border. 'Thank God,' the peasant cried with relief, 'now I won't have to endure any more Russian winters!'

We may smile at the peasant's belief that Russian weather will halt at the border like a thwarted army, but attitudes not too dissimilar to that of the peasant have been found in psychological studies. For example, in one study people were less concerned about the threat of contamination from a proposed nuclear power plant that might be built a given distance from their home if there was a political border between their house and the plant.[22] And yet, presumably, nuclear contaminants have no more respect for borders than Russian snow. It's therefore perhaps not surprising that, when we chop up continuous scales into categories, using arbitrary boundaries (like degree classifications), our own perception of reality can be similarly distorted. And that can lead to some very unfortunate effects.

Take weight – it's an obsession in the developed world. In the USA alone, the weight-loss industry was worth an estimated $64 billion in 2014. One way of estimating whether a person's weight is healthy is to calculate their body mass index (or BMI), which is their weight in kilograms divided by the square of their height in metres. While the BMI is a continuous scale, the World Health Organization (WHO) divides it into categories – below 18.5 is interpreted as signifying that a person is underweight, 25 to 29.9 suggests they are overweight, and 30 or more indicates they are obese. However, many researchers have used different cut-off points without a clear rationale.[23] Before 1998, the National Institutes of Health (NIH) in the USA used a threshold of 27.8 for men and 27.3 for women before they were considered to be overweight. When this was changed to 25, for both genders, to conform to the WHO guidelines, millions of Americans who'd believed they had a normal weight suddenly found they were labelled as 'overweight'.[24] This might seem like an innocuous change – after all the Americans in question had not become heavier overnight. However, as we shall see, the placement of boundaries between categories can have an effect that is anything but neutral.

In an experiment, two psychologists, Francesco Foroni of SISSA in Trieste in Italy and Myron Rothbart of the University of Oregon in the USA, presented participants with silhouette drawings of nine female body types which appeared on a continuum from very thin to very heavy.[25] In the first phase of the experiment participants were asked to assess the similarity between pairs of silhouettes. After a ten-minute break, arbitrary boundaries were introduced – the three thinnest silhouettes were labelled as 'anorexic', the middle three were labelled as 'normal', and the three stoutest, labelled as 'obese'. The participants were then asked to repeat the similarity judgements. This time they perceived that the silhouettes sharing the same label were more similar than those having different labels. In other words, two adjacent silhouettes were judged to be more similar if they belonged to the same category than if they were separated by the arbitrary boundary. The effect persisted even when the legitimacy of the boundaries was challenged and even after the boundaries were removed and the similarity judgements repeated.

Similar boundary effects have been found in other studies. People asked to estimate the difference in average temperature in Providence, Rhode Island, between dates eight days apart (e.g., 19 and 27 June) expected larger differences when the interval crossed a monthly boundary (e.g., 27 June and 5 July).[26] Once a boundary is established, our perceptions overestimate the true similarities within a category and exaggerate the differences between categories. The effect appears to survive arbitrary changes in the boundaries. When job applicants' aptitude scores were divided into 'ideal, 'acceptable' and 'marginal', and the cut-off points between the categories were moved, people's assessment of the similarity of the applicants was still greater when they were in the same category than when they appeared in different ones. This was the case, despite the assessors being told that the boundaries were purely arbitrary.[27]

All of this matters because labels can be surprisingly powerful in their effects on the way we see ourselves and others. In a longitudinal study, girls labelled as 'too fat' in childhood were found to have higher chances of an 'obese' BMI nearly ten years later, irrespective of what their BMI as children had been.[28] The researchers speculated that stress, resulting from the labelling, had caused the girls to resort to coping strategies like overeating. There is also evidence that labels can affect women's body image to the extent that they may have implications for mental health.[29] In another study, teachers at an elementary school were told by researchers, Robert Rosenthal and Lenore Jacobson, that certain children had achieved marks on an academic test that put them in the top 20 per cent of achievers.[30] These children were labelled as 'academic bloomers', but in reality they had been randomly selected and their abilities typically were no different from other children at the school. Yet, when the researchers returned a year later and retested the children, the 'bloomers' scored 10 to 15 IQ points more than their peers. The label had become a self-fulfilling prophecy: perhaps because the teachers had devoted their efforts to nurturing the intellectual development of the 'bloomers'. It's easy to infer from this that a child's future could be heavily dependent on which side of an arbitrary boundary line their mark falls when taking an examination or aptitude test.

So why do we love to chop up numeric scales?

We've seen that chopping up numeric scales into cruder categories leads to a loss of information and can cause people to have a distorted perception of reality. Why do we do it? After all, years ago the Irish dramatist George Bernard Shaw said: 'Crude classifications and false generalisations are the curse of organised life.' Since the time of the ancient Greeks, philosophers have been troubled by the issue of where to place boundaries on categories. The sorites paradox (a name derived from *soros*, the Greek word for a 'heap') refers to a situation where an individual grain of sand is taken from a heap. It is argued that the original pile of sand would still remain a heap. But what if we carried on successively removing single grains? The same argument suggests that each removal would still leave the pile as a heap. Eventually we would end up with just one grain, but surely this could no longer be called a heap. At some point in the process, the collection of grains must have transitioned from belonging to the category 'heap' to the category 'non-heap'. But when was the boundary crossed? When did the tiny change – the removal of a grain of sand – lead to the big switch between categories? The same issue can apply to continuous scales such as temperature or height, where infinitely small changes are possible as opposed to changes that come in irreducible quantities, like whole grains of sand. At what temperature does a room cease to be warm? At what height does a person become tall?

Despite these philosophical challenges, categories help us to cope with a complex world that assails us daily with a deluge of information. Brilliant though the human mind is, with its billions of neurons and synapses, it still has severe constraints on the amount of information it can process at one time. Replacing a continuous scale with a few categories reduces the amount of information we have to process. Categorisation also allows us to make sense of the world. There are continuous scales designed to measure a person's degree of introversion or extroversion, but it's much easier to explain people's behaviour at a party by allocating them to binary categories: he's an introvert and she's an extrovert. It's likely that this urge to categorise evolved to ensure our survival. Our ancestors may have lived in an environment where there were many species of snakes, some of them harmless. But assigning all snakes to the category 'Dangerous Creatures!' would be likely to be advisable when they suddenly encountered a new species hanging from a tree deep in the forest.

Communicating with others is also simplified if we use categories.[31] Colours appear on a continuous electromagnetic spectrum, but I might have difficulty attracting someone's attention if I shout: 'Just look at those beautiful 472 nanometre wavelength flowers over there!' Replacing '472 nanometres wavelength' with the word 'blue' would probably work much better. But blue is still a category with arbitrary boundaries: wavelengths of 450 to 495 nanometres are often used to delimit the colour blue, but these can vary. Of course, my ability to assess colours to an exact wavelength would be problematical anyway, unless I carried a spectrophotometer around all the time.

So chopping up scales is natural and can be helpful. But we should also be aware of the dangers and distortion that this can lead to: the illusory step changes, like cliff edges, that we see between categories and the potentially dangerous stereotyping of those who fall between the imaginary boundaries. In the reporting of scientific studies, these thresholds can allow a small change in a measure to be magically transformed from an 'insignificant result' to a 'significant finding' or vice versa. They can enhance or diminish a graduate's or schoolchild's future prospects and they can even have serious implications for people's mental health and self-image. Boundaries may not mean much in themselves, but if we accept them unthinkingly, their effects can be insidious and far-reaching.

5

A Daily Life in Numbers

Unreal steps

Something annoying happened to me in Santander, Spain, one lovely June evening a year or so ago. Santander has a broad tiled promenade that leads in the west towards heady clifftop paths with superb views along the coast, so a long stroll seemed a perfect way of spending the last few hours of the day. I must have walked several miles enjoying the warmth and the soothing rhythm of crashing waves. It wasn't until I returned to my hotel that I realised that a minor misfortune had occurred – I hadn't been wearing my fitness tracker. All those steps and all that effort expended in climbing to the top of the cliffs had gone to waste. According to the tracker, I'd failed to hit my daily steps target. It was almost dark, but for a second I entertained a mad thought: I could do the walk again. The light lingers long into the night in June and I was sure I could find my way to the summit of those cliffs again.

Fortunately, sanity prevailed. I joked that the steps I'd taken weren't real ones because there was no record to them, but I still felt cheated. I recalled an article I'd read in the *Guardian* that featured a woman who confessed: 'Recently, I have started to worry: do I even exist without my Fitbit? Without data, am I dead?'[1] I wondered whether I was suffering from a similar syndrome to tourists who only see the world though a camera lens – or a smartphone screen – because the objective is to record rather than to experience. Or perhaps maximising some number had supplanted living, feeling and enjoying. I had a friend whose aim was to visit as many countries as he could. He had a list of nations accompanied by long rows of ticks that he constantly wanted to extend. Once, he had taken the boat from Cape Town to Tristan da Cunha – a six-day voyage – but had stepped ashore for the shortest possible time. All he wanted was the landing stamp in his passport.

It got me thinking. To what extent is the richness of our human experience impoverished when we focus on quantifying it or achieving numeric goals? Does the truth about our lives reside in recorded numbers, uncontaminated by our unreliable memories and psychological vagaries? Or is there something more?

Lifelogging

In recent years there has been an exponential rise in the availability of new technology that enables us to track many aspects of our lives. You can wear or swallow devices that will record your geographical location, physical activity, sleep quality, heart rate, brain waves, calories burned, quality of social interactions (using properties of your voice), mood, and level of happiness. There are devices like the i.Con, a smart condom developed by British Condoms in 2019, that will quantify and record aspects of sexual activity such as its duration, number of thrusts, calories consumed, and the range of different positions used over a specified period. Some devices even record the volume of noise produced by participants and the numbers of partners they have had over time. Users can join league tables so they can compare and advertise their sexual prowess.[2] Even babies in the womb, and those yet to be conceived, are not exempt from quantification. The latest wearable technology will monitor a prospective mother's resting pulse rate, skin temperature and breathing rate in order to advise on the best time to conceive. After conception, other devices will track the foetus's heart rate and number of kicks, together with the mother's sleep position.[3]

Of course, this urge to track and document all the details of one's path through life is not new; there have always been people who have kept diaries. Some have used alternative formats. In 1726, Benjamin Franklin, who was then 20 years old, began to keep a daily record of the extent to which he achieved thirteen personal virtues, including temperance, silence, frugality, sincerity and humility. He focused on each virtue in turn for thirteen weeks before moving to the next one. Over 250 years later, in 1998, Gordon Bell, a computer scientist at Microsoft, decided to keep an electronic record of a huge number of aspects of his life, including all his correspondence, books he had read, meetings, conversations and phone calls. In 2003 he began to wear a camera, the size of cigarette packet, which automatically takes a photograph approximately every 20 seconds, as long as it is not moving. Although he does not wear the camera all the time, by 2012 he had recorded a hundred sequences, consisting of 60,000 to 80,000 images. These include pictures of walks to conferences, meals and people he has spoken too. He calls his project 'MyLifeBits'. Then there is the Italian conceptual artist, Alberto Frigo, who, starting in 2004 when he was 24, has been photographing every object his right hand uses. On average, he takes 76 photographs a day and he plans to continue until 2040, by when he estimates he will have amassed a million pictures. The images he's gathered so far have been displayed on huge walls in art exhibitions.

These examples don't involve recording lives in numbers, but the availability of the latest hi-tech gadgets has led to a trend towards quantification. There is now a movement called the Quantified Self – a term created in 2007 by two *Wired* magazine editors, Gary Wolf and Kevin Kelly. By 2016 the Quantified Self Institute had over 70,000 members with meet-up groups established in cities around the world. The movement's motto is 'self-knowledge through numbers'. Wolf himself reported a typical record of his day:

> I got up at 6.20 this morning, after going to bed at 12.40 a.m. I woke up twice during the night. My heart rate was 61 beats per

> minute, and my blood pressure, averaged over three measurements, was 127/74. My mood was a 4 on a scale of 5. My exercise time in the last 24 hours was 0 minutes, and my maximum heart rate during exercise was not calculated. I consumed 400 milligrams of caffeine and 0 ounces of alcohol. And in case you were wondering, my narcissism score [based on 'a well-validated psychological test that takes only a few minutes'] is 0.31.[4]

Even dry numbers like these have been turned into art. From 2005 to 2014 the American graphic designer Nicholas Felton produced and sold glossy books containing annual reports of his self-tracked data, displayed in aesthetically pleasing charts and numbers.

Lifelogging these days is not only about monitoring our own bodies and personal experiences. You can extend it to track the quality of relationships with others: the Quantified Relationship has arrived.[5] For example, the app Kouply can turn your relationship with your partner into a game. You can reward each other points for a gift of unexpected flowers, a foot rub, a passionate kiss, or cooking a perfect romantic meal. Couples who choose to can join a leaderboard to compete with others. Billed as 'the app that encourages couples to be grateful and playful for a strong, lasting relationship', its Seattle-based creators, who work at Microsoft, hope that one day marriage counsellors will prescribe their product.[6]

There appear to be several motives for tracking your daily life or relationship in numbers, besides knowing yourself. Perhaps the most common is a desire for self-betterment. 'Unless something can be measured, it cannot be improved,' says Kevin Kelly, the grey-bearded co-founder of the Quantified Self movement. Some people, like the University of Canberra professor Deborah Lupton, see this motive as being consistent with the contemporary zeitgeist, in which individuals, rather than the state, are deemed responsible for their own health and happiness.[7] In this perspective, a failure to achieve these objectives reflects on the individual, taking no account of the possibility that they may be socially or physically disadvantaged. Other sociologists and psychologists argue that the last sixty years or so has seen a decline in conformity to traditional patterns of living, so individuals now have a range of choices on how to shape their lives. But this choice also brings uncertainty. Against this background, the self-tracking devices come with the promise that you can take control and maximise your life chances.[8] You can become an 'optimised human being'. The main threats to us no longer come from bacteria, viruses or natural disasters, argues Roger Taylor, who co-founded the health analysis firm Dr Foster Intelligence. They arise from our inability to control ourselves.[9]

Competition can be an effective stimulus to self-improvement, and, as we've seen, many of the tracking devices on offer allow users to compare their performance with others in league tables (the ugly word 'gamification' is sometimes used to describe this process). But this also means that achieving status can be another motive for self-tracking. For some people this seems to pre-empt a desire for self-improvement and can even lead to dishonesty. The *Wall Street Journal* reported that people have confessed to attaching their activity trackers to hamster wheels, power tools, ceiling fans and their pet dogs.[10] A year or so ago, in one UK league table in which participants posted their monthly step counts, one user was claiming that they were walking 5 million steps a month – much to the anger of those who could not claim the number one spot, despite their aching feet and thighs. That's nearly 100 miles a day, month after month. Their other data suggested that they were active for 23 hours a day. Perhaps they had a hyperactive pet.

Feelings versus figures

Some Quantified Selfers advocate total faith in numbers as an accurate record of one's life. If I don't feel stressed, but the number displayed on my tracker tells me I am, I should believe the tracker. Our feelings, judgement and memory, it is argued, are unreliable. Much research in psychology supports these arguments when they relate to judgement and memory. Amos Tversky and Daniel

Kahneman became world famous for their work demonstrating the biases in our judgement, likening them to optical illusions. Our recall of the past is distorted by recent or unusual events. We see correlations between different phenomena that do not exist and we perceive illusory patterns in random events. Reality is viewed through a lens that is warped by our current mood and emotions. 'Computers, don't lie. People lie,' argued Monica Hesse in a *Washington Post* column in 2008.

In contrast, numeric measurements are emotionally detached, precise, and free of our psychological vagaries and limitations. They bring the comforting promise of certainty. By analysing them, it is argued, we can discover patterns and correlations that we would otherwise be unaware of. Perhaps my app reveals that my happiness level always drops on a Friday morning; I can start to explore why. Or maybe the hours I'm awake during the night are correlated with my blood pressure levels the following day, so I should explore ways to improve my sleep. Even better, if individual statistics are pooled into vast data sets, created when people upload the readings from their devices, valuable discoveries about what causes what are possible.

Larry Smarr, a physicist at the University of California, San Diego, wanted to monitor his weight and diet. Out of shape and pre-diabetic, he enrolled in a personal blood-testing service to track the ratio of Omega-6 to Omega-3 fatty acids in his blood. The tests produced a large amount of data, and, among these, he found that the readings for one variable, C-Reactive protein, were consistently anomalous. This led him to undertake further tests, and eventually, to his surprise, he found that he was suffering from Crohn's disease. Speaking at a conference in 2012, he said: 'This idea that you can just feel what is going on inside of you, that is just so epistemologically false. You just can't do it.'[11] Nowadays, Smarr tracks and graphs a hundred variables based on blood tests that are taken quarterly or annually.

To dedicated Quantified Selfers, the body is seen as a machine, providing performance readings like a car engine. Some people even speak of 'a dashboard for the body'. When happiness, moods, and the quality of social interactions are included, perhaps 'a dashboard for daily life' would be a more appropriate term. But there are critics of this 'reductionist' trend. 'Get a life!' suggested Dr Rowan Williams, former Archbishop of Canterbury, speaking on BBC Radio 4. Arguably, many aspects of our individuality and well-being are excluded from the array of measurements on the dashboard. Some are missed because they cannot be measured, or at least not easily measured; others because self-tracking technologies reflect the concerns and values of their designers, predominantly 'white, well-paid, heterosexual men living in the Global North'.[12] For example, the Apple Watch did not include a menstrual-cycle tracker when it was first launched.

Some people fear that the measurements may be selected by app designers primarily because they produce information that will be profitable when sold to third parties. Insurance companies have shown a keen interest in accessing data generated by users and several have signed deals with app developers.[13] Others worry that the designers have an eye to maximising sales or stimulating continued engagement (with the app – not your intended), rather than improving well-being or the quality of romantic relationships.[14] As such, they might favour measures that are easy to understand or based on popular, but unproven, beliefs about health. For example, few studies have found that calorie-counting apps help you to lose weight.[15]

The very act of measurement itself may lead to distortion and create its own reality. On the Quantified Self website, one self-tracker, Jeff Kaufman, reported his experiences of tracking his happiness for a year. His phone pinged at random times and asked him to record his happiness level on a scale of 1 to 10. Instead of thinking 'How do I feel right now?', he wrote, it's hard not to just think 'In past situations like this I've put down 6 so I should put down 6 now.' But he adds: 'Being honest to myself like this can also make me less happy.'[16] Others worry that obsessive measurement of health variables can turn people into hypochondriacs. Or it can itself create unhappiness, by making people feel that they are under constant pressure to improve. 'One pound heavier this morning? You're fat. ... Skipped a day of running? You're lazy,' wrote one former self-tracker, explaining why she had given up the pursuit so that the tool wouldn't be 'an instrument of torture anymore'.[17] Lifeloggers may even perceive a lack of control over their lives – the very opposite of what self-tracking is intended to achieve.

Even if every aspect of our lives could be accurately measured, the argument goes, the whole may be greater or less than the sum of the parts. Feeling healthy is more than a bundle of medical and physical quantities. Fitness is not the same as the number of steps you've walked in a week. Sexual enjoyment cannot be represented by a formula that combines numbers of thrusts, noise levels and duration. Apps that measure calories burned during sex might even encourage people to pursue the wrong goals – sex becomes a means to fitness, an alternative to the gym, rather than as a pleasurable end in itself. Deborah Lupton points out that numbers may look scientific, neutral and objective, but the algorithms that combine them in some products to produce aggregate scores are hidden away in 'black boxes'. We may therefore have no idea what assumptions they make when they perform their calculations and give more or less emphasis to one number rather than another. For example, one app combines measurements of breathing and body movements to give you a quality of sleep score on a 1 to 100 scale. Another combines sixteen nutrition and exercise measures to give users a weekly health score. With self-tracking, these neat measurements replace the untidiness and richness of our lives, whose complexities are thus reduced to a stream of lifeless bytes. Critics such as the Belarussian writer and thinker Evgeny Morozov claim that there is a danger that our trust in our subjective experience will decrease in favour of what the bytes tell us: there is, he argues, an 'imperialistic streak' in quantification.[18]

Numbers as a health supplement

In reality, the idea of self-trackers as a homogeneous group of narcissistic, number-obsessed nerds whose lives have been reduced to digits, graphs and dashboards is almost certainly untrue. Although this may characterise a few people, the evidence is that self-trackers are a varied bunch and for many the numbers are merely supplemental to their life. When Tamar Sharon of Maastricht University and Dorien Zandbergen of the University of Amsterdam conducted an in-depth study of self-trackers, they found that they had a range of motives, but most regarded their intuitive feelings and the numbers as complementary.[19]

For these people, it was a combination of tracked data and intuition that helped them towards the self-knowledge or improvements they aimed for. After all, the numbers don't speak for themselves: they have to be interpreted and that involves bringing one's subjective perspective to bear on them. One person they interviewed called this 'digital storytelling' – the numbers simply enhanced the narrative you were telling yourself or others. As the researchers argued: 'Quantified data helped [their participants to] render aspects of a private, subjective and somewhat inaccessible world of feelings and problems more tangible and comparable.'

Some self-trackers in the study even claimed that the tracking process helped them to develop a sixth sense or a greater awareness of the world. One person, who had been tracking his food intake for some time, found that he had developed the ability to determine intuitively how many calories were in a portion of food and what its weight was. Another said that, after using the tracking devices for long enough, you no longer need them – you can sense intuitively what they would have told you. For people like these, the daily stream of numbers is a secondary consideration. They merely serve as a means towards developing a more intense awareness of life through cultivated senses.

Even a conflict between the numbers and one's feelings might prove to be enlightening. In decision making there is evidence that, when your intuition tells you one thing and the numbers tell you another, it pays to investigate why the two are opposed.[20] You might be buying a new car: your calculations tell you that a small hatchback is your best choice, but your intuition is bidding for an SUV. The very process of investigating the discrepancy can lead to new insights and clarifications so that you arrive at a better, more informed, decision. It may reveal that your intuition is flawed or that the numbers were misleading or incomplete. Sense is balanced against sensibility. In a similar way, it's possible that probing such conflicts in self-tracking can yield new insights. My tracker is telling me that I slept well with several hours of deep sleep and REM sleep, but I feel tired and headachy. What's the explanation? Perhaps the bedroom was insufficiently ventilated, or I was dehydrated.

Romance and love seem the least likely candidates for quantification. Choosing a future spouse by rating candidates on their attributes and choosing the one with the highest overall score would, in most people's eyes, be the antithesis of romance. But, for those who have already become a couple, the Quantified Relationship has its defenders. One criticism is that these apps, which record and score what each person has done for the other, turn the relationship into a cold economic exchange. After all, each partner is valued for what they give the other. John Danaher of the National University of Ireland, Galway, and his fellow authors point out that in many relationships there is an imbalance in who carries out household tasks and other chores, and that in heterosexual relationships this is often to the detriment of the woman.[21] An app might highlight these imbalances, and for some couples, the authors argue, this could lead to a fairer distribution of chores and a better love life. Anyway, they point out, relationships already exist both as means to an end, such as financial security, and for their own intrinsic value, and there is no reason why one motivation should preclude the other. Quantifying some aspects of a relationship can still allow for spontaneity and informality.

So is self-tracking a curse or a blessing? In an ideal world trackers would provide useful supplementary information, helping to guide our decisions as we muddle through life, potentially enhancing our health and well-being. But we need to acknowledge that the numbers they create are just low-hanging fruit. While they are easy to obtain, they can only provide a partial, and possibly biased, impression of our daily lives. They are a passive and pale reflection of our experiences, and if we aren't careful they will narrow our vision and obscure more important, less tangible truths.

This can lead to pointless obsessions. I've just been pipped for first place in a league table for the number of steps walked in a month. I hate not winning, so I'll make sure that I top the table next month. My resting heart rate has mysteriously increased by four beats a minute. I'd better phone the doctor. My body is telling me that I need to rest this evening, but I've not met my target for calories burned. I'd better head for the gym. Trackers may be promoted as devices for improving our lives, but they can make us feel pretty fed up, too.

6

Polls Apart

Shocking surveys

'98 Per cent Demand Ban New Migrants', declared the front-page headline of the *Daily Express* on 1 November 2013, below a modest claim that it is: 'The World's Greatest Newspaper'. '80 Per cent Want to Quit the EU', the same paper told us on its front page on 21 January 2015. 'British youngsters "most illiterate" in developed world', reported ITV News on 29 January 2016. 'Shock Poll: Most Countries Prefer Putin over Trump', Forbes.com revealed on 16 August 2017.

All these shock findings were based on surveys. But surveys are fragile barometers of reality: one weakness in their design or implementation can leave their needles pointing hopelessly in the wrong direction. Even the most carefully constructed ones are likely to be only approximately right. But when you want the needle to point to an extreme end of the scale – to support your argument or manufacture an eye-catching headline – it's best to ride roughshod over careful construction; malfunctioning parts are especially welcome.

Take the report in 2015 that 80 per cent of Britons want to quit the EU. The *Daily Express* claimed that the result was based on 'the biggest poll in forty years'. That sounds impressive, but the 80 per cent finding seems high when just over a year later, in an official referendum, only 51.9 per cent of people voted to leave the EU. So who were the people polled? It turns out that two Conservative MPs and a parliamentary candidate had delivered 100,000 forms to households in their three adjacent constituencies in Northamptonshire. This had taken them several months, but despite their efforts, only 14,581 completed forms had been returned to the *Daily Express* offices.[1] No effort had been made to poll a representative cross section of the public throughout the UK, and, on top of this, there was a serious danger of self-selection bias – the 'Disgusted of Tunbridge Wells' phenomenon (or, in this case, the Disgusted of Wellingborough). Just as angry people are more likely to write to newspapers, so those with very strong opinions will be more motivated to respond to polls like this, but their views probably won't be fully representative of those of the general population.

The 80 per cent figure must have been disappointingly low for the *Daily Express* editors. In other polls they have found that: '99 per cent of you say get us out of Europe', '99 per cent say Britain should not spend more on foreign aid', '99 per cent think that beggar gangs should be sent back to Romania', though admittedly only '97 per cent think we should stop paying out benefits for feckless families'.[2] Unlike the MPs' poll, these astonishing figures come from so-called voodoo polls – a deprecatory term coined by the well-known pollster, Sir Robert Worcester, to describe polls where TV viewers or newspaper readers are invited to vote on different options with no controls being put in place to obtain a representative sample of opinion. In the *Daily Express* polls, people vote by dialling one of two premium-rate phone numbers, so the poll is driven purely by self-selection from a pool of readers who are likely to be sympathetic to the newspaper's political orientation. There is also nothing to stop a zealot voting multiple times, as long as they can afford the phone bill. The result is that the percentages that shout at you in a font that's big enough to take up a third of the front page have absolutely no statistical validity at all.

Distorted frames

Respectable polling organisations make every reasonable effort to obtain accurate estimates of facts about populations, but they can't survey everyone. Only governments have the resources for such large-scale exercises. The US 2010 census cost $13 billion (about $42 per person),[3] while the 2011 census in the UK cost around £482 million (over £7 per person) and employed 35,000 people.[4] Even then, accurate results are not guaranteed. America's 1990 census overlooked an estimated 8 million people, predominantly immigrants and minority ethnic groups. Ten years later around 17 million people were counted twice.[5] When the 2001 UK census lost about a million young people, the official responsible said he thought they were probably all in Ibiza.[6] The colossal logistics involved in taking a census also means that it takes a long time to get answers. It's no good announcing the results of an opinion poll eighteen months after an election. And TV executives won't want to wait two years before finding that viewing figures for a show have crashed.

Usually, the only practical way to get timely information about a large population is to take what we hope is a representative sample. (In quality control in industries where items are tested to destruction there is no other option than taking a sample. If I manufacture matches and decide to test them all by striking them, I won't be in business for long.) But how can we try to ensure that a sample is an accurate reflection of the entire population? To start with, it helps to have a list, or a database, containing details of all of the members of the population (this is called a sampling frame). But this is often easier said than done. I would struggle to find a complete list of beer drinkers, Oxford Street shoppers, vegans, or people who will definitely vote in the next UK election, and any available lists might be grossly unrepresentative.

An unrepresentative sampling frame was the main cause of one of the most famous statistical disasters of all time: the 1936 *Literary Digest* fiasco. In that year, Franklin D. Roosevelt and Alf Landon went head-to-head in the US presidential election. Roosevelt, the incumbent, was a Democrat who in his first term had launched the

New Deal policies to counter the effects of the Great Depression by creating employment and bringing relief to the poor. Landon, the Republican candidate, was governor of Kansas. Although he was sympathetic to some aspects of the New Deal, he objected to what he called the 'bungling and waste of social security' and he argued that Roosevelt's policy had had a minimal effect on unemployment. As in previous elections, the *Literary Digest*, a respected weekly magazine, decided to conduct an opinion poll in which it contacted 10 million Americans – a huge sample by any standards. It told its readers that they would know the election result in advance 'within a fraction of 1 per cent of the actual popular vote'. Over 2 million people returned their mock ballot forms and the magazine confidently predicted that Landon would win a landslide victory with 57 per cent of the vote, carrying 32 of the 48 states. In the event, Roosevelt crushed his opponent with 61 per cent of the vote. Only two states – Maine and Vermont – gave Landon a majority. It was one of the most one-sided elections in US history. The post-election issue of the *Digest* had four words on its cover: 'Is our face red!'

So why did the *Literary Digest* poll get it so wrong? First, there was the obvious possibility that the 20 per cent who bothered to respond to the survey may have been more likely to support Landon than those who ignored it – Disgusted of Wellingborough transferred to an American setting – giving a skewed impression of the political landscape. But, even if all 10 million had participated, it's unlikely that the poll would have been accurate. The magazine selected its respondents from its own readers, telephone directories and car registration lists. People on these lists in 1930s America were relatively wealthy – only one in four households had a phone – and, as such, they had the least to gain from Roosevelt's policies and would be more inclined to vote for Landon. Despite the huge effort and expense associated with the poll, its unrepresentative frames meant that it was doomed from the outset. At least Alf Landon, who lived to be 100, gained some consolation from the historic defeat and the infamous election forecast. 'They might have forgotten me if it had been close,' he observed.

Who's picked for the sample?

If a pollster can obtain a sampling frame that accurately lists a population, without significant omissions or duplication, they can choose to select a random (or probability) sample. Essentially, this involves assigning a number to every individual on the list, and then, in the manner of a raffle, randomly picking a selection of these numbers.[7] This means we can calculate the probability of any individual in the population being selected for the sample: if 2000 people out of a population of 1 million are to be surveyed, each person has a 2 in 1000 chance of being in the sample. Because everyone has the same chance of being included in the survey, there is no systematic bias towards selecting particular groups of people, such as those who are rich, male or university educated.

In practice, rather than shaking hats containing numbers written on folded pieces of paper, the selection is usually carried out by a computer which has been programmed to generate random numbers. I used to think that these numbers came from a roulette wheel spinning somewhere in the mysterious depths of the computer, but in fact most computers can only produce pseudo-random numbers.[8] These are numbers that look random, but which can be predicted if you know the algorithm that produced them. For example, an old-fashioned method, proposed by the mathematician and computer scientist John von Neumann, involves starting with a so-called seed number, squaring it, and then treating the middle digits as the random number. If a seed is 38, squaring it gives 1444, so the number 44 has been generated, and if you are number 44 on the list you have been selected for the survey. We then square 44 to get 1936 so that our next 'random' number is 93, and so on. The method has serious limitations, however. Once 00, 50 or 60 have been generated, the method will get stuck on that number (for example, $50^2 = 2500$). If the method produces the number 24 it will keep returning to that number every two iterations ($24^2 = 576$ and $57^2 = 3249$, so the sequence would be 24, 57, 24).

Over the years much more sophisticated methods have been developed, but none are truly foolproof and some widely used software products, such as Microsoft Excel, still use outdated random number generators. This means that we can't be sure that everyone on the list will have an equal chance of being selected. (Despite this, Canada was recently using Excel to 'randomly' select 10,000 people for permanent residency status from a list of 100,000 applicants.[9])

Even if the sampling process is truly random, so that there is no systematic bias, there's still no guarantee that a representative cross section of the population will be selected in a particular sample. By chance, almost every person who comes out of the metaphorical hat could be rich or male or someone in their twenties, despite the population consisting of all income groups, both genders and all ages. However, if we have information about the structure of a population, we can increase the sample's reliability through what is referred to as stratification (the result is called a stratified random sample). For example, if 55 per cent of the population is female, to select a sample of 100 people we could put all the female numbers in one hat (so to speak) and the male numbers in another. By randomly selecting 55 numbers from the first hat and 45 from the second, we would ensure that our sample reflects the make-up of the population, at least in terms of gender. The problem is that, to do this, we don't just need a list of people's names before we start sampling, we also need to know their gender (and possibly other information such as their age and social class). Sometimes this information can be collected as part of the survey and the results subsequently adjusted (or reweighted) to compensate for groups which may be under- or over-represented in the samples.[10]

Rather than picking people from a sampling frame and then trying to contact them, it's often easier to confine the survey to people who are instantly accessible because they happen to be on the street or at home when the survey is being conducted. In this case, the interviewer, rather than chance, determines whether a person should be included in the sample. To try to ensure that interviewers don't confine themselves to particular types of people, polling companies usually set quotas of people of different ages, genders and social classes who will be interviewed (the method is called quota sampling). As with stratified sampling, the sizes of these quotas are intended to reflect the make-up of the target population. For example, in 2017 it was estimated that about 11 per cent of the British adult population were females aged over 65, so in a sample of 1000 people you would aim to interview 110 females in this age group. However, because the choice of who to survey is left to the interviewer, we no longer have a random sample with everyone having the same chance of being polled. Nor can we calculate the chances of any individual being questioned. Interviewers are more likely to stop people in the street who look friendly or appear not to be in a hurry. Shopaholics are more likely to appear in a high-street survey than those who despise a trip to the shops. Some groups of people, such as shift workers and teenagers, will be less likely to be at home if an interviewer calls at their house at a particular time.

There are tales of interviewers panicking when filling their quota was proving to be impossible and driving up and down streets in a desperate bid to find someone – anyone – who fitted the quota's specifications. Buck Buchanan, an interviewer for Gallup in the US in the early days of quota sampling, recalls the temptation to fudge a person's age or income so that they could be included in an incomplete quota. Sometimes interviewers simply made them up because they were too embarrassed to ask about such personal details. Then there were the people who fitted the quota but were hostile. Buchanan recalls a farmer responding to a question: 'That ain't none of your business!' When he persisted, the farmer pointed his gun. 'Now, you git! And don't come around here asking who I'm going to vote for! Git.' Buchanan marked the man down as a Republican voter.[11]

These days much opinion research is carried out on the internet. You don't have to pay interviewers to stand in the street or knock relentlessly on the doors of people who resent having their favourite television programme interrupted. (Others may be delighted by the visit. When a former neighbour of mine was working as an interviewer, an elderly lady answered the door. 'Come in,' she smiled. 'It's the first time I've had a man in here for five years.") And unlike a telephone survey, you'll be avoiding the expense of thousands of wasted calls where suspicious voices at the other end of the line utter expletives before slamming the receiver down. Many Americans were so fed up with calls asking for their political opinions that, in the summer of 2015, Gallup agreed to pay out $12 million to settle a class action lawsuit brought on behalf of all those who'd received unsolicited calls from the company between 2009 and 2013. (Gallup denied any wrongdoing.)

However, internet surveying is not without its problems. Usually there is no frame available listing the internet users relevant to your sample. Nor is there a practical way of randomly generating email addresses, and even if there were, many people would regard the unsolicited messages as spam. This means that pollsters have to resort to other methods and these have the potential to generate misleading results.

A common method is the opt-in panel. Internet users are recruited, often via websites, to become members of panels to take part in future surveys. The polling organisation then selects random samples of its panellists for particular surveys. Usually, panel members are rewarded for their responses. For example, in the UK the Opinium agency invites people to 'sign up and join over 40,000 members all getting paid to have their say on the day's talking points'. In the US the Harris Poll urges people to 'become a member and start earning rewards for your favorite brands'. While the selection of respondents for a given survey is random, the panel itself is, of course, not a random sample of the general population. Those attracted to panels tend to be younger, more computer literate, and less well-off.[12] In political polling those responding are usually more politically engaged.[13] There are also known to be 'professional' respondents who sign up for multiple panels and have a tendency to rush through the questions with scant attention in order to claim their reward quickly.[14] Of course, being young, computer literate, or politically engaged does not necessarily mean that a person's response to a particular issue will be untypical of people in general. Where panellists don't tend to differ in their opinion from the general population, opt-in panels can work as well as random samples, especially if adjustments are made to take into account their make-up.[15] Also, random samples can suffer when some people don't respond to the survey. There has been a catastrophic decline in response rates over recent years when random telephone surveys are used. One US study found typical response rates declined from 36 per cent in 1997 to just 9 per cent in 2013.[16] Those who choose to respond to a random-sample survey are, like panel members, self-selecting.[17]

Mythical margins of error

In theory, random samples have a key advantage over non-random samples. Because we know the chance of anyone in the population being selected for the sample, we can calculate its reliability. For example, we might be able to say that our opinion poll result, based on a random sample of 1000 people, has a margin of error of plus or minus 3 per cent, meaning that we are almost certain, usually 95 per cent certain, that the true figure lies within 3 percentage points of our estimate. The bigger the sample, the less the margin of error: if 2000 people are polled, this might be plus or minus 2 per cent. We can't perform this calculation for non-random samples, so we have little idea what the survey's reliability is. In quota sampling, the whims of an interviewer in choosing who to interview defy mathematical measurement. With opt-in panels, we don't know the probability that someone in the population will be motivated to join the panel. Despite this, pollsters commonly report the margins of errors of polls even when the survey is non-random. For example, Public Policy Polling in January 2019 indicated that 46 per cent of voters were in favour of impeaching President Trump with a margin of error of plus or minus 3.6 per cent.[18]

In practice, the margins of error quoted for *both* random and non-random samples are mythical, and in general, they underestimate the amount of uncertainty associated with a poll result. For random samples, incomplete and inaccurate sampling frames and non-responders add to the uncertainty about how reliable the results are, but these are not factored into the calculation. When a poll's results are broken down into subgroups, such as men and women or those aged under 21 and those aged over 50, they will be based on smaller samples, so any margin of error associated with them will be relatively large. These bigger margins of error are rarely published. The mathematics also tells us that the estimated margin of error should depend on the percentage of people giving a particular response, so the results for different questions will vary in their margin of error. When 50 per cent of people say 'Yes, Trump should be impeached' and 50 per cent say 'No, he shouldn't', the margin of error is highest because this is when there is the greatest uncertainty about the balance of opinion in the population. If only 5 per cent say yes and 95 per cent say no, the estimated margin of error is more than halved. Yet many pollsters report a single margin of error for the entire survey. Typically, this assumes that 50 per cent of respondents have given a particular response, so, ironically, this factor on its own can lead to the margins of error for many questions being overestimated.[19] For non-random samples, stated margins of error are even less trustworthy. They not only suffer from all of the above issues, but they are also based on the brazenly false assumption that the sample is random.

Worryingly, there's yet another doubtful practice that can make the polls seem more reliable than they really are. During elections, when the pollsters feel the glare of publicity on their backs, a big gap between their findings and those of their rivals can make them feel exposed. There's a risk of being the pollster who got it badly wrong when everyone else was close to the actual election result. This motivates some pollsters to adjust their figures so that they closely match those of the other polls, a phenomenon known as herding. The result can be an illusion of accuracy: if all the polls are giving us the same message, surely they must be tapping into some underlying truth. There's a strong suspicion that herding took place in the weeks before Australia's 2019 federal election. The polls had been within 0.25 percentage points of each other and all had indicated that Bill Shorten's Labor Party would defeat the existing Liberal–National coalition government led by Scott Morrison. But, as Kevin Bonham, an independent analyst from Tasmania, pointed out, there is so much uncertainty in election polling that it's 'vastly improbable' that all the polls would be so close.[20] After all, the polls themselves would claim a margin of error of 2 or 3 percentage points. In the end, Morrison's shock victory meant that all the polls suffered a crisis of credibility and many commentators and politicians called for them to be scrapped.

It looks like a certainty

'Poll gives big lead to Heller' was a headline in the *Las Vegas Review Journal* in October 2008. It was referring to a forthcoming election in Nevada where a poll suggested that the Republican Dean Heller would get 51 per cent of the vote, while his Democrat rival, Jill Derby, would only get 38 per cent. But the poll was unusually small – only 221 registered voters were questioned – and the report estimated that the margin of error was plus or minus 7 percentage points. As we've seen this margin of error is likely to have been underestimated, but if it's to be believed, Derby might have been the choice of as many as 45 per cent of voters, while Heller's support could be as low as 44 per cent, so Derby might actually be in the lead. This possibility is increased if the poll's true margin of error is larger.

The problem is that the uncertainty surrounding polling results doesn't make the headlines and this creates a false impression of exactitude. Journalists often run stories that highlight movements in the polls. On 5 September 2018, the *Independent* newspaper reported: 'The survey puts Jeremy Corbyn's [Labour] party at 41 per cent – up from 40 per cent in July.' The Labour leader might have been heartened by the result, but the 1 percentage point increase was probably just noise – a chance variation that falls well within the margin of error. If you take two samples, they'll usually give slightly different results, even if opinion in the population remains unchanged, simply because different people will have been questioned. Only large changes are unlikely to be due to chance variation.

Separate studies in Sweden and Demark found that newspapers often compounded illusions of change in public opinion by providing political explanations for small, potentially random, movements in the polls.[21] 'Candidate A's lacklustre performance in last week's debate has caused a 2 percentage point decline in his lead.' 'People are warming to Candidate B's promise to reduce income tax. It's led to a 3 percentage point increase in her poll rating.' We have an intolerance of randomness, so the media obligingly overlays it with narratives, and this doesn't just apply to opinion polls. The financial pages provide stories for the daily twists and turns in a stock market graph: a poor harvest in Brazil or expectation of a US interest rate increase. Sports pundits eagerly explain what's suddenly gone wrong with the Manchester United team after they narrowly lose three football games in a row. Yet, according to Chris Anderson, a professor of economics and politics at Warwick University, 'Football is driven more by luck and chance than other team sports.'[22]

We can't totally blame the media for turning meaningless movements in the polls into stories or for paying scant attention to the uncertainty of their findings. As we'll see in Chapter 9, colourful stories have the edge over dry numbers when it comes to attracting our interest. And stories are better supported by exact-sounding single numbers than vague ranges. 'Labour pulls ahead in poll by 4 percentage points' is a headline that might hook us into reading an article. 'Poll indicates with 95 per cent confidence that Labour's lead is somewhere between minus 1 and plus 9 percentage points' will probably encourage us to turn to the next page. Single numbers are easier to absorb in a world where masses of information compete for our attention.

Despite this, when we are aware of uncertainty there is evidence that we will tolerate estimates presented as ranges as long as they aren't too wide, acknowledging that an exact number is not appropriate.[23] However, once a range widens beyond a certain point, our tolerance of it tends to diminish. Overly wide ranges are regarded as uninformative, even if they accurately represent the amount of uncertainty involved. In one study, people were given estimates, expressed as ranges, of the number of countries that were members of the United Nations. Ninety per cent of participants thought the range 140 to 150 was a better estimate than the wider range 50 to 300, even when they were told that the correct figure was 159.[24] Given that real margins of error in polls, if they could be calculated, would be likely to produce wide ranges, most people would probably prefer to believe the headline figure.

Does it matter if journalists and their readers treat polling figures as exact and authentic reflections of the public mood? There's plenty of evidence that it does. Polls can influence people's opinions, shape government policies and agendas, and have a direct effect on voters' behaviour. Benjamin Toff of the University of Minnesota found in two experiments that some people have a tendency to conform to views that the polls suggest are held by the majority of the population.[25] The number of people supporting taxes on junk food was 9 percentage points higher among those who had seen a poll result indicating that this policy had majority support (the poll result was actually made up for the purpose of the experiment).

In elections, polls can produce bandwagon effects when a party's upward movement or a narrow lead increases its support among previously uncommitted voters. Many countries have seen such surges in support, including Britain, Holland, France, Austria, Denmark and Germany.[26] Tod Rogers of Harvard University and Don Moore of the University of California, Berkeley, claim that polls can also have an underdog effect.[27] This benefits candidates who are narrowly behind in the polls, because it galvanises the efforts of committed supporters who see the small gap as bridgeable. This may have been to Donald Trump's advantage in the 2016

US presidential election. In the run-up to the vote he was usually behind Hillary Clinton in the national polls, but her lead was often as low as 1 per cent. Trump clearly liked to be regarded as the underdog. At a rally in Miami just under a week before election day, he told his supporters: 'The polls are all saying we're going to win Florida. Don't believe it, don't believe it ... Pretend we're slightly behind ...'

Broken barometers?

In recent years pollsters have taken a hammering. Shocks on election nights now seem the norm. Hillary Clinton surely expected to wake up on the 9 November 2016 to find that she would be the first female president of the USA. David Cameron was not expecting to win Britain's election in 2015 – he thought that he'd never have to implement his pre-election pledge to hold a referendum on EU membership – but he did win. And then there was the surprise of the referendum result itself, followed soon after by the astonishment that Theresa May had lost her majority in the 2017 election. In France, *Le Parisien* decided not to publish any polls during the 2017 presidential campaign in the light of their inconsistent performance in the primary elections. In Brazil, the polls seriously underestimated the support for Jair Bolsonaro of the far-right PSL party that won the 2018 presidential contest. In Australia in 2019, as we've seen, there was widespread incredulity when the government was re-elected.

Yet, here's another surprise. When Will Jennings of the University of Southampton and Christopher Wlezien of the University of Texas at Austin analysed more than 26,000 polls from 338 elections that took place in 45 countries between 1942 and 2013, they found that there was no evidence that the typical accuracy of polls has got worse. If anything, it has marginally improved.[28] There is, they argue, nothing to justify the perception of a crisis in polling – a few high-profile disasters doesn't mean that the whole polling industry is in chaos.

But that, of course, doesn't mean that today's polls are precisely accurate and that we should take their headline figures as givens. It only tells us that today's polls are just about as good or bad as those of eighty years ago, and anyway the findings only apply to election polls. So when a report informs us that 74 per cent of people surveyed in England say they have used cocaine at least once in their lives compared with 43 per cent globally and that Britons get drunk on average 51 times a year compared to a global average of 33, we need to question who was surveyed and how they were selected, and remember that even the best-designed surveys can be well off-target.[29] Polling data is intoxicating stuff: drink too deeply of it and one can end up looking rather foolish.

Even if a sample turns out to be a near-perfect representation of a population, it's still wise to be sceptical about figures like these. Skewed samples are not the only factor that can lead the findings of polls and surveys astray. As we'll see in the next chapter, when it comes to questions about lifestyle factors, like drinking or drugs, and many other aspects of our lives, many of us are prepared to lie.

7

On a Scale of 1 to 10, How Do You Feel?

Street opinions

The middle-aged male canvasser stopped me in the street with the deft footwork of a premier league full-back. Cornered next to the plate-glass window of the local stationer, I didn't have time to think of reasons why I couldn't be interviewed.

But then there followed a torrent of questions about my attitude to this or that proposed government policy, how I rated half a dozen politicians as prospective prime ministers, and countless other topics. I sensed the interviewer's intelligent, unblinking eyes scrutinising me over half-moon glasses as I blustered my instant responses on issues as complex as climate change and international aid. A 'four' for him, a 'five' for her, loads of 'yeses', the odd 'give that five: strongly agree', an occasional, simple 'that will be a one: strongly disagree', but never 'please can you give me an hour to research and think about that difficult question before I respond'.

Sometimes, the same questions seemed to return, but phrased slightly differently. My goodness, I'll look a fool if I give a different answer, I thought, but what did I answer last time? Occasionally, none of the options listed by the interviewer quite fitted the response that I would have liked to have given, but I chose one of them nevertheless. Then came the questions about personal interests. Seventies pop music and football wouldn't chime with this interviewer, I surmised, so I restricted my answers to baroque music, broadsheet cryptic crossword puzzles, and international travel (but not to theme parks, I added). But only 'travel' had a box on the interviewer's list, so he scribbled my additional answers under 'Others, please state'.

I left the interviewer feeling slightly embarrassed and guilty that I'd probably misrepresented myself. I'd felt pressurised to express strong opinions about things I'd thought little about, and my aim had been solely to persuade this complete stranger that I was an intelligent rational being and to reassure him that he'd picked a good interviewee. I suspected that other people may have felt the same. Yet these days more and more data is being collected on our subjective responses either by personal interview or in mail questionnaires or internet surveys.[1] Often we are forced to express these responses on numeric scales so that warm emotions and rich and complex perceptions have to be translated into cold, sparse numbers.

Of course, ever since money was invented, we've had to make subjective judgements about numbers. Is that pig worth 50 groats or that bag of potatoes worth 2 florins? But now numeric scales have extended their range into traditionally qualitative territory, such as happiness, pain, strength of preference, intensity of agreement with a proposition, quality of experience, competence, and artistic merit. How satisfied are you with your life on a scale from 1 (awful) to 5 (wonderful)? To what extent do you think that racism is prevalent in society on a scale from 0 to 10? How painful was your endoscopy? Rate your stay at the River Hotel from one star to five. How many days of your life would you be prepared to give up to have your back pain cured? And the subjective judgements demanded by questions like these and others can make or break businesses, influence medical practice, determine where charitable donations should be channelled, lead to employees being fired or promoted, determine whether or not you pass an examination, and even inform the policies of governments. So can they be trusted?

Questionable answers

Back in the 1950s, an American organisation called the Color Research Institute was worried about seemingly unreliable responses to its surveys, so it decided to conduct a number of experiments. In one of these, it surveyed people with the question: 'Have you ever borrowed money from a personal loan company?' One hundred per cent of the respondents said no. Apparently, some of them even came close to shouting the answer. Yet all of the interviewees had been selected because they were listed as clients of local loan companies.[2] More recently, studies have found that around 25 per cent of non-voters in American elections report that they have voted when asked immediately after an election.[3]

These days, market researchers and others who conduct surveys are aware that people, either deliberately or inadvertently, give untruthful answers to surveys. The problem is that it's difficult to estimate the extent of untruthfulness or to filter it out. One consequence was the disastrous performance of the opinion polls in Britain's 1992 general election. The Conservative party, led by John Major, was deeply unpopular and had been shaken by a series of sex scandals. As a result, so-called 'shy Tories' were reluctant to admit to pollsters that they were intending to vote Conservative. On the eve of the election, the polls gave the opposition Labour party a 1 per cent lead, inspiring an exuberant performance by its leader, Neil Kinnock, at a rally in Sheffield. Kinnock's cheerfulness proved to be ill-founded. The next day the Conservatives held on to office with a 7.6 per cent lead over Labour.

Moulding one's answers to conform to what is judged to be socially acceptable is even more likely when lifestyle issues, such as sexual behaviour, drinking habits, or consumer choices, are being explored in surveys. Dr Terri Fisher, a psychologist at Ohio State University, surveyed 293 students, ostensibly to find out how frequently they engaged in behaviours that were regarded in the prevailing culture as 'typically male' or 'typically female'.[4] For example, telling obscene jokes and yelling at other drivers were seen as masculine behaviour, while lying about your weight and writing poetry were seen as predominantly female traits. Four further questions asked the students about their sex lives – for example, the total number of partners with whom they had had sex. Fisher then randomly divided the participants into two roughly equal groups. Both groups were asked to answer the questions, but the students in one group were wired up to a realistic-looking lie detector which in reality was fake. Responses to questions relating to stereotypical gender behaviour were unaffected by the presence of the fake lie detector. For example, women in both groups were equally willing to report that they engaged in what was regarded as 'typically male' behaviour, like wearing dirty clothes. But, when it came to reporting sexual activity, the men who were not wired up to the lie detector reported having more sexual partners than their counterparts in the other group. The women, on the other hand, reported having fewer sexual partners when the fake lie detector was not present. It seems that, when they thought they could get away with it, both sexes were prepared to lie about their sexual behaviour to match the cultural expectations of their gender. Fisher concluded: 'There is something specific to reports of sexual behavior that leads participants to respond in gender-stereotyped ways.'

Alcohol consumption is another area where people tend to be 'economical with the truth'. Estimates of drink sales obtained from surveys of the population tend to be far less than actual sales as measured by beverage companies or tax authorities. Partly, this is because heavy drinkers may refuse to participate in surveys in the first place, but people who do respond appear to have a poor recall of how much they have drunk in the recent past. An Australian study suggested that, on average, people underestimate their consumption by 40 to 50 per cent.[5] Young males and middle-aged females produced the biggest underestimates.

Then there are consumer choices. The economists Maria Loureiro and Justus Lotade conducted a survey in supermarkets in Colorado, asking people how much extra they would be prepared to pay for coffee products that were labelled as 'fair trade' or were produced in an environmentally friendly way.[6] People questioned by an African male interviewer indicated they would be prepared to pay a significantly higher premium than those interviewed by a white American male. Presumably, the respondents wanted to please the African interviewer, since fair trade and reduced environmental damage would benefit people living in the interviewer's continent. There are plenty of other examples of people's stated preferences for products or services belying their actual choices because they want to be regarded in a favourable light, either to themselves or an interviewer. In various studies, respondents overstated their preferences for non-genetically-modified food products,[7] electric cars,[8] local and organic apples,[9] goods that are packaged in an environmentally friendly way,[10] milk and meat where the welfare of livestock could be verified,[11] and shopping in local rural stores.[12]

Even an interviewer's style of dress can influence responses. Visitors to a forest were asked, hypothetically, how much they would be willing to pay annually to support a woodland conservation scheme. When the interviewer wore 'a well-tailored navy blue business suit, white full length shirt, tie, and black leather shoes' they were generally willing to pay more than when he wore 'a T-shirt, knee-length shorts, and white trainers'.[13] The researchers emphasised that all the interviewer's items of clothing were clean and well pressed throughout the survey.

Of course, answering hypothetical questions in a survey is not the same as making real decisions about what to buy or how much to pay. There are unlikely to be consequences if you give a wrong answer in a survey, so people may devote little mental effort to deciding how to respond. The American researchers Jon A. Krosnick and Duane. F. Alwin presented people with a list of thirteen desirable qualities that a child might possess, such as having good manners or getting along with other children. They then asked people to indicate the three qualities that they considered to be most important. Even though the list was presented in a different order each time, qualities that appeared near the top were significantly more likely to be selected as most important.[14]

Sometimes people give erroneous responses because they have forgotten the answer. For example, can you recall how much you spent on electricity and water last year or what your usual monthly expenditure has been, over the last three months, in restaurants and cafes? Both of these questions have appeared in American surveys.[15] Our memories are far from perfect. When the Dutch psychologist Willem Albert Wagenaar carried out an experiment on himself in the 1980s, he found that 20 per cent of critical details of events in his life could not be retrieved from memory one year later. And that's despite the fact that he had specified that these details were 'certain to be remembered' at the time of the event.[16]

On other occasions people may simply not know the answer to a question, but when they feel pressurised by an interviewer, they may be loath to admit it. It's long been known that some people are even willing to give their opinions on fictitious issues or non-existent countries. For example, in an early experiment in the 1940s, many respondents blithely expressed negative views about the people of Wallonia.[17] In a later American study, people were questioned as follows: 'Some people feel that the 1975 *Public Affairs Act* should be repealed – do you agree or disagree with this idea?' Around a third of those questioned indicated whether or not they agreed when they were pressed for a response, yet no such act appeared on the US statute book.[18] Similarly, in a British study, 10 to 15 per cent of the public gave their opinions (on a five-point scale) on the Monetary Control Bill and the Agricultural Trade Bill, neither of which existed.[19] People who indicated they had an interest in politics were most likely to express a view on the acts, presumably because they thought that a politically aware person ought to have an opinion.

Know thyself

One indication that people are unsure about their opinions is 'lability' – a condition where an individual's responses fluctuate wildly over relatively short periods of time. When Americans were asked to indicate on a numeric scale the extent to which the USA should cooperate with Russia or get tougher with that country, only about half gave the same answer when they were questioned again six months later.[20] In an experiment I conducted with Semco Jahanbin and other colleagues at the University of Bath, we provided students with information and pictures of different types of consumer goods, such as mobile phones, laptop computers and television sets, and in each case, asked them which product they would prefer to buy if they had to make a choice.[21] We then asked the same people the same question on two further occasions, each two months apart. For most of the products, there were significant changes in their indicated preferences for different brands over the three surveys. Of course, to some extent this may have reflected changes in technology or the publication of news items and product reviews about the brands between the surveys. It may also have resulted because the choices were not being made for real, so the participants did not think that it was worth devoting mental effort to thinking hard about their decisions. But some of the inconsistency may have arisen simply because the respondents didn't really know what their true preferences were.

'The most difficult thing in life is to know yourself.' The quote is attributed to Thales, the ancient Greek philosopher who is regarded by many as the father of science. Knowing your own preferences and attitudes may reflect this difficulty, especially when asked to indicate what you would prefer in a hypothetical situation. I might happily indicate to an interviewer that I'd enjoy parachuting out of an aeroplane and be confident that I have given an honest response, but when the real prospect of leaping into a ten-thousand-feet void is presented, my enthusiasm might be a little dimmed. And I might eagerly tell a market researcher that I'll definitely buy that stylish four-door electric car that will be launched on the market next month. But when the following month arrives, the reality of parting with nearly £25,000 hits home and I decide to stick with my existing five-year-old petrol model.

Some researchers argue that in many situations most of us don't have a set of true preferences or attitudes anyway. Rather, we have in our heads a ragbag of half-formed and partly consistent thoughts and ideas. When asked for our opinion, we are prompted by the phrasing of the question, the immediate context, and by recent experience to bring to mind a few of these scraps of thoughts which we then use to help us choose between the options.[22] As a result, we may give different responses when questioned on a particular issue on different occasions. The ordering of questions in a survey can influence which thoughts and ideas come to mind and so determine how we respond. In one well-known study, students at the University of Illinois were asked to indicate their satisfaction with life on a scale ranging from 'not so happy' to 'extremely happy'.[23] However, one group of students were asked to answer a preliminary question first: 'How happy are you with your dating?' Another group only answered the dating question *after* they had indicated their satisfaction with life. When the dating question came first, the answers to the two questions were significantly correlated. Those who tended to be happy with their dating tended to say they were more satisfied with their lives and vice versa. But when the dating question came second, the correlation disappeared. It seems that asking about dating caused particular thoughts to surface and hence affected how the students answered the life satisfaction question. A similar result was found when researchers asked Americans how satisfied they were in general with their local government services.[24] If they were first asked to rate specific public services, such as parks, police, schools and transportation, they expressed lower general satisfaction than if the general question preceded the question about specific services.

When a numeric response is required, there is always the danger of planting a number in people's minds which then influences their answer. Psychologists call the planted figure an anchor, and once it's in your mind, it's difficult to make a big adjustment to it when giving your response. For example, if a person is first asked 'Would you pay more or less than £50 for this product?' and is then asked

'How much would you pay for it?', the £50 acts as an anchor and their response to the second question is likely to be close to this figure. If, instead, we first asked 'Would you pay more or less than £25 for this product?', we would probably get a lower response to the second question.

Then there's the wording of survey questions themselves. Subtle changes in the way questions are phrased can have big effects on people's answers. Many Americans are sceptical about climate change despite the fact that an overwhelming majority of scientists are convinced that it is occurring. This motivated Jonathon Schuldt, then a PhD student at the University of Michigan, and two of his professors to conduct an experiment.[25] They suspected that the terms 'global warming' and 'climate change' could evoke different attitudes to the issue. For example, they argued that 'global warming' calls into mind temperature *increases*, which may seem to run counter to headlines about record snowfall or reports that New York is having its coldest day for ten years. In contrast, 'climate change' may be seen as relating to temperature *changes* and these are more easily associated with the snowmageddons that sometimes bring cities to a halt at unexpected times of the year. Also, they had evidence from other research that, while 'global warning' conjured up connotations of human impact on the climate, natural causes tended to be more implicated in 'climate change'. The researchers noted that, while the term 'climate change' predominates on liberal think tank websites, conservative think tanks tend to refer to 'global warming'. They suspected that this was not accidental. For example, it may be easier for conservatives to find arguments that attempt to discredit 'global warming' as opposed to 'climate change'.

To test the effect of using these different terms when eliciting the American public's views, they asked over 2000 people to respond to the following question:

> You may have heard about the idea that the world's temperature may have been going up [changing] over the past 100 years,

> a phenomenon sometimes called global warming [climate change]. What is your personal opinion regarding whether or not this has been happening?

Roughly half the people received the term in brackets rather than the preceding term. All the respondents were asked to indicate their opinion on a scale from 1 (definitely has not been happening) to 7 (definitely has been happening). As the researchers expected, significantly more respondents chose a value of 5 or above, which implied that they believed the phenomenon was happening, when the question referred to 'climate change' rather than 'global warming'. When the results were analysed at a deeper level, it was found that this difference was largely driven by Republican voters. For many of them, the simple transposition of the two terms in the question made a significant difference.

It's not only the words that can influence our answers to questions. When asked to indicate a response on a numeric scale, the way the scale is presented can have significant effects. In a study in the 1990s, a representative sample of German adults were asked how successful they had been in life.[26] A total of 34 per cent chose values between 0 and 5 when the scale ran from 0 (not at all successful) to 10 (extremely successful), suggesting that they did not rate their success too highly. But, when the scale ran from –5 (not at all successful) to +5 (extremely successful), only 13 per cent chose values between –5 and 0. It appears that the respondents thought that a negative number represented explicit failure in their life, while the low numbers on the 0 to 10 scale represented an absence of success, which they were more willing to acknowledge.

Happiness as a number

So how happy are you? How happy is the country you live in? In the last twenty years or so, happiness research has exploded and armies of pollsters, clipboards at the ready, are roaming the world trying to get answers to questions like these. Here's a typical question they might pose:

> Please imagine a ladder with steps numbered from zero at the bottom to ten at the top. Suppose we say that the top of the ladder represents the best possible life for you and the bottom of the ladder represents the worst possible life for you. On which step of the ladder do you feel you personally stand at the present time?[27]

And here are typical outcomes, based on sets of questions like these.[28] Finland, according to the 2018 World Happiness Survey, is the happiest country in the world with a score of 7.632 (yet its suicide and homicide rates are among the highest in the Western world and alcohol abuse is the leading cause of death of Finnish men). It nudges Norway into second place, with a score of 7.594. The USA languishes in 18th position (score: 6.886), one place ahead of the UK (6.814). Right at the bottom of the table is the landlocked African state of Burundi – its score is only 2.905.

Nowadays, governments are using results like these to forge policies. Since 1971 the Himalayan kingdom of Bhutan, a country of breathtaking scenery and ancient Buddhist monasteries, has been using Gross National Happiness, rather than Gross Domestic Product, to gauge progress. The UK has followed, belatedly, with its government now regularly measuring personal well-being at the national, regional and local levels. It does this by asking questions such as: 'Overall, how satisfied are you with your life nowadays?' and 'Overall, to what extent do you feel the things you do in your life are worthwhile?' People are asked to respond on a scale of 0 to 10, where 0 means 'not at all' and 10 means 'completely'. The results are presented with great precision. For example, the 2018 report highlighted that in Scotland the average rating for the feeling that things done in life are worthwhile had improved from 7.81 to 7.88 between March 2017 and March 2018.[29]

But can we really measure happiness, the happiness of an entire nation, in a single number presented to two or three decimal places? Personally, I have difficulty working out where I am on the imaginary ladder, with its eleven rungs, or where I sit on the UK's 0 to 10 scale. If I'm forced to commit myself, then recent, superficial and salient events are likely to dominate my thinking – a rude email, a friendly chat with a neighbour, a sunny day, a bad cold, or even a good cup of coffee will send me up and down the ladder like a window cleaner.

I suspect I am not alone in this. The psychologist Norbert Schwarz has found that, when asked to make a judgement such as how happy we are, how beautiful or risky something is, or how truthful a statement is, we use our immediate feelings and moods to guide our assessment, even when these are irrelevant to what we are judging.[30] For example, we may rate a passage that is easier to read as being more truthful than one that we struggle with. Similarly, we may a consider a food additive that is harder to pronounce, like Fluthracnip, to be more dangerous than one that is easier to say, like Magnalroxate.[31] In one widely reported study, which Schwarz conducted as a PhD student at Mannheim University, he asked people to indicate their life satisfaction in a questionnaire. Before they began, he asked them to photocopy a sheet of paper. Half the respondents were lucky enough to find a small coin – planted by Schwarz – on the photocopying machine. The temporary uplift in their mood, caused by the apparent chance discovery, was sufficient for them to rate their entire lives as being more satisfying.

As we saw earlier, the nature of the question and its position within a questionnaire can have significant effects on people's answers. For example, the ladder question refers to the 'best possible life for you'. This may evoke notions of what an ideal, and probably unattainable, life might look like. Because most of us will see our lives falling short of the ideal, the question itself may cause us to lower our assessment of how happy we are.[32] This is especially likely because psychologists have shown that the misery we feel if we are worse than a reference point is more intense than the joy we would experience if we were better than it.[33] People in relatively poor countries may watch programmes like *Friends*, which depict people living a wealthy middle-class lifestyle in the USA, and conclude that they have a long way to go up the ladder even though they may be quite happy in themselves.

Even asking a question about personal happiness might have a negative effect. 'Ask yourself whether you are happy, and you cease to be so,' said John Stuart Mill. This is particularly true when the question calls attention to factors that a person normally keeps at the back of their mind. Observations of the behaviour of people with serious disabilities, such as paraplegia, suggests that they can enjoy life as much as the general population.[34] Once they have adapted to their situation, they no longer spend all their waking hours thinking about it; instead they get on with the rest of their lives. But, when a question reminds them of their condition, their response is likely to be a gloomier assessment.

Yet another problem is the definition of happiness, especially when making international comparisons. Terms like 'satisfaction with life' and 'well-being' are often used interchangeably with happiness, but may represent different things. For example, some researchers think of well-being in terms of the attainment of pleasure and the avoidance of pain. Others see it as a reflection of the extent to which a person has a sense of meaning and self-realisation in their life. Even the simple word 'happy' does not have exact translations in other languages. As the Danish economist Christian Bjørnskov points out, both the French and Russian translations mean happy *and* lucky, while the Danish translation *lykkelig* is a stronger notion than 'happy' in English. Nevertheless, Denmark regularly comes at the top, or near to the top, in international league tables of happiness.

So should the success of governments be judged on their nation's position in such league tables? And should government policies be aimed at improving happiness scores, rather than GDP? Advocates of happiness measures, such as Lord Richard Layard of the London School of Economics, argue that they should. Although self-assessments of happiness suffer from the limitations we have discussed, Layard argues that they still correlate well with objective measures such as people's cortisol levels (cortisol is a hormone that the body releases mainly in times of stress). He also points out that people used to mock the idea that you could measure levels of depression, but now its measurement is universally accepted. In addition, measuring variations in happiness between people, he argues, will allow us to find out which factors lead to happiness and which do not.[35]

Others argue that depression and happiness are different and independent psychological states and not simply opposite ends of a spectrum. While levels of certain hormones and other biological markers can predict depression, their association with happiness is less clear. It is not yet possible to validate self-assessments of happiness against objective measurements.[36] The philosopher Julian Baggini has more fundamental concerns and worries about the very idea of trying to measure happiness.[37] What makes life worthwhile, he argues, is not necessarily what makes you happy or satisfied – it's more complex than that. Life can seem worthwhile when you are dissatisfied with your current position and relish the challenge of making improvements: gaining new qualifications or skills, becoming physically fit, or securing a more interesting and worthwhile job. If we can't accurately define the good life, he argues, we may be measuring the wrong things.

Happiness researchers have tried to address some of their critics by exploring new measurement methods. For example, experience sampling aims to overcome the danger that we will assess our entire satisfaction with life based on our feelings in a single moment by asking us to respond to questions when our mobile phone bleeps at a series of random times during the day.[38] Attempts to test the validity of self-reports of happiness include recording people's frequency of smiling or laughing or asking other people to rate how happy they think a person is and seeing whether these correlate with how happy people say they are. However, without a clear understanding of what happiness is, it's difficult to argue that happiness measurement is anything other than a meretricious attempt to assign quantities, via unanswerable questions, to a nebulous concept, without any possibility of objective calibration. Those three-decimal-place differences between the happiness of different nations might provide entertainment for newspaper readers and confer an aura of scientific exactness on the results, but closer reflection reveals that they are a classic example of spurious accuracy – micrometre readings of an ill-defined phenomenon reflected in error-bound data.

Pain on the scales

Having a baby can be one of the happiest events in people's lives, but for some women the pain of labour is about as bad as it gets. They report intense cramps with internal twisting as if someone is trying to pull their organs out, and at least one mother has likened the experience to being run over by a train. Yet, in the 1940s, several women allowed a white-coated scientist to stand over them as they gave birth and, between their contractions, calmly burn their hands and ask them how it felt. The scientists were based at Cornell University and they were seeking to create a measure of pain intensity that they called a 'dol', after *dolor*, the Latin for 'pain'; they called their pain measurement techniques dolorimetry. Their scale ran from 0 to the maximum pain that can be experienced by a person, 10.5 dols.[39]

The research team was led by the physicist James D. Hardy, an energetic Texan who was a veteran of the Normandy campaign in the Second World War. The team's intentions were entirely worthy. If pain could be measured, they reasoned, it would help doctors to assess the effectiveness of analgesics and other pain-reduction treatments. Thirteen women agreed to take part in the study; 'either curiosity or a desire to be of service caused them to volunteer readily'. However, during labour, when thermal radiation was being administered to their hands to a point beyond where their skin was burned and blistered, a Mrs O. 'soon began to show hostility to the entire team in spite of expressing many times prior to admission her desire to take part in the study'. A Mrs U. also 'cried and complained with vigour' even though her pain measurements 'indicated only 2 to 4 dols'. The other women apparently cooperated. One participant, a woman who had previously experienced six miscarriages, allowed the team to administer pain of 10.5 dols that caused second-degree burns. The researchers reported that 'she wished to cooperate fully as an expression of her gratitude in having a term pregnancy and insisted on having the tests made'.

Happily, the women were not required to express their pain level as a number on the 0 to 10.5 scale. Instead, they reported whether the sensation on the back of their hand was more or less intense than the most recent uterine contraction they had experienced. Depending on this report, the burning was either increased or decreased, allowing the scientists to estimate a numeric pain level that was approximately equivalent to that caused by the contraction. Typically, this required three or four burnings, each of three seconds.

Despite the heroics of these women and the medical students who agreed to have their foreheads burned in similar experiments, the methods pioneered by Hardy and his colleagues did not catch on. Other scientists found that they could not reproduce their results. In particular, the type of pain experienced in a controlled scientific setting did not appear to be the same as 'real' pain. The participants in the experiments conducted by Harvey's team had, to some extent, been trained. They were willing to take part, informed about what to expect (so the pain was, in part, predictable), and they were able to develop an attitude that shielded them from distracting mental influences that might have altered their experience of pain. In subsequent years, pain came to be regarded as a distinctive personal experience that varies between individuals, rather than a phenomenon that could be standardised.[40] For example, there is some evidence that redheads and people who are non-athletic, significantly overweight or depressed have a lower tolerance of pain.[41] Sensitivity to pain even varies within individuals – most people can tolerate more pain on the dominant side of their bodies (e.g., the right side if you are right-handed).[42]

To be of value, pain-assessment methods therefore face considerable challenges. They need to be reliable so that they will elicit consistent responses from people when applied under similar circumstances. They also need to measure what they are supposed to be measuring – for example, they shouldn't be reflecting a person's anxiety, their mood, or their reluctance to report pain because they don't want to appear troublesome to medical staff. Older people, in particular, are less likely to report pain because they were brought up in an era when complaining about pain was regarded as a sign of weakness or because they think pain is part of the normal process of ageing.[43] Lastly, a method should be easy for people to understand. People in pain will not want to battle with complicated questions and they are unlikely to give reliable answers if they are confronted with such questions.

Recent research suggests that asking people to report their pain on a 0 to 10 scale comes closest to meeting the above challenges – at least for adults or those who are not cognitively impaired.[44] This method also appears to be preferred by patients. Scales that contain words such as 'mild', 'moderate', 'severe' and 'very severe' can cause people to cluster their responses around the position of the words so that small changes in pain over time may be less easy to discern. These changes may also be more difficult to detect when a scale simply shows a ruler with two endpoints, typically labelled as 'no pain' and 'worst imaginable pain' and a person has to mark the point on the otherwise featureless line to represent their pain. However, facial expressions representing different levels of pain can be useful where children are involved.

So it seems that people are generally willing to translate the pain they are suffering into a number and that this can provide medical staff with useful information when managing their treatment. Of course, we cannot know whether the number they choose is anything more than a rough expression of what they are subjectively experiencing. Because of this, some scientists, aided by advances in areas such as brain imaging and biomarker research, are seeking the gold standard of objective pain measurement. If successful, this could be used to validate, or even replace, the patient self-reports that we currently have to rely on. This aim is proving to be challenging. For example, levels of stress hormones such as cortisol and adrenalin are influenced by other factors besides pain, while levels of sweating and skin conductance can be affected by a person's skin quality or the ambient temperature.[45] Nevertheless, in 2013, American scientists used advanced computer algorithms to process the images of over a hundred brains of people who had been exposed to heat that ranged from being pleasantly warm to painfully hot.[46] They found common patterns in the images that signified the level of pain being experienced. Interestingly, they also found that physical pain differs in the response it generates in the brain from social pain, such as the distress a person experiences after the break-up of a relationship (an earlier study had yielded brain images of people who had just been shown pictures of the person who had rejected them).

The American study only analysed the experience of moderate pain applied to the forearm of young healthy volunteers. Extreme pain, chronic pain, pain in other parts of the body, or pain experienced by tired, frightened or sick people might produce very different patterns in the brain. This research is therefore only in its early stages and some people may reject the very notion that pain can be measured objectively anyway.[47] But, if it succeeds, it will not only be a boon to the medical profession, it will also tell us whether those subjective numbers that doctors have used for years have been accurate.

Charitable decisions

At the end of September 2018, a magnitude 7.5 earthquake struck the Indonesian island of Sulawesi. It triggered a tsunami that swept ashore with waves of up to twenty feet, destroying homes, offices, hotels, mosques, and transport and communication links. Near the devastated city of Palu, houses were carried for more than a mile in a torrent of mud, while ships and boats were tossed like toys on to the land. Thousands of people were killed, injured, or left to sleep in the open amid the desolation. Aid for the survivors was slow in arriving – they lacked food, water and shelter, and many were forced to loot supermarkets to stay alive. Around the world, television programmes broadcast desperate appeals for international help. In Britain, the Disasters Emergency Committee (DEC) asked the public to contribute to the aid effort. Like millions of others, I found it impossible to resist such pleas.

Shortly afterwards I read a book that suggested that I should resist the inclination to donate when natural disasters strike.[48] Donating purely because you have an emotional reaction to tragic events was a mistake, it told me. Instead, the book's author, William MacAskill, an associate professor of philosophy at Oxford University, argues that I should be making a rational decision which would involve me researching where my money would be likely to do the most good in improving the world. Charitable donations are usually subject to the law of diminishing returns. The benefits of the early sums received are greater than the later sums because the most basic and crucial needs are met first. Each extra £1 therefore brings fewer and fewer added benefits. Because millions of other people would be responding to the disaster appeal, the marginal benefits of my contribution to Indonesia would be relatively small, so my money would bring bigger gains if I donated it elsewhere. In particular, there are ongoing tragedies that don't make the headlines, such as the 18,000 children who die each day from preventable causes. I should have sent my money to help to alleviate these tragedies, the book argues, by donating to cost-effective charities.

MacAskill is a leading proponent of effective altruism – 'a do-gooder who uses his head more than his heart', according to the *Sunday Times*. While doing the most good with your money sounds sensible, it does require that you are able to assess the benefits of your gift. Would a donation of £500 to help a blind person to have a guide dog lead to more benefits than the same donation to an eye clinic in east Africa? Or should I donate to charities that help wounded army veterans, or prevent child abuse, or support cancer research?

Advocates of effective altruism recommend the use of a measure called the QALY, or quality-adjusted life year, a measure that is also recommended for decisions about the allocation of health resources.[49] A QALY of 1 is a year of life in perfect health, so if I can expect to live another ten years in perfect health, that's 10 QALYs. However, if those ten years will be marked by continuous lower back pain, then my quality of life might be rated as only 75 per cent of that of a perfectly healthy person, so I have only 7.5 QALYs left. Being dead implies a QALY of zero – but states worse than death could exist and would be reflected in a negative QALY. According to effective altruism, if my £500 donation would increase a Congolese man's QALY by 0.5 or a Cambodian woman's by 0.8, then I should donate to the Cambodian female – assuming I have only £500 to give. Improvements in QALYs can be aggregated for whole groups of people. For example, one visit of a mobile eye clinic to a remote region might be estimated to improve the QALYs of each of 5000 people by 0.4 on average, yielding 5000 × 0.4 = 2000 extra quality adjusted life years. If this costs £40,000, that's £20 per QALY – a measure of how cost-effective the intervention has been. But how can QALYs be assessed? Who is to say that having lower back pain means that one year in a person's life only has 75 per cent of the quality experienced by a perfectly healthy individual?

A number of methods have been used. One method is to ask people to respond to a question like the following:

> Imagine that you will live for another ten years. Suppose that you can either (A) live these ten years with lower back pain or (B) live another eight years in full health. Which would you choose?

The person might respond that they would choose the eight years. As the objective is to find a point where the person is indifferent between options A and B, we would need to modify option B to make it less attractive. We could therefore repeat the question but offering instead six years of extra life in option B. This time they might say that they'd prefer the ten years with lower back pain. Suppose that, eventually, when option B offers seven years of life in perfect health, the person says that they would be indifferent between A and B. This suggests that, for this person, ten years with back pain equals seven years of perfect health. So a year with back pain is equal to 0.7 QALYs.

Of course, this figure only applies to the individual who was questioned, so responses are averaged across patients or representative cross sections of the public who have been surveyed. Averages that have been elicited by researchers include 0.9 for mild angina (so a year with mild angina is equivalent to 0.9 years in full health), 0.45 for being anxious, depressed and lonely much of the time, and 0.99 for a life with menopausal symptoms.[50]

Personally, I have difficulty in answering the lower back pain question, for a simple reason: I'm lucky enough not to suffer from the condition. This means that I have to imagine what it would be like before I can decide whether to choose A or B. This raises the issue: should QALYs be elicited from the general public or from patients suffering from the specific health condition? In the UK, for example, it's the general population who provide the data. It's argued that, since they don't have a vested interest in attracting resources to particular areas of medicine, they are less likely to give biased answers.[51] But, as we've seen, people often imagine that the experience of some health conditions is worse than the reality because people adapt to the condition and are still able to enjoy their lives. Indeed, their quality of life may be as good as that of a perfectly healthy person. QALYs treat the absence of illness and quality of life as being the same.

There are many other challenges for the assessment of QALYs. Some people may refuse even to contemplate the idea that they might shorten their lives to ameliorate a health condition. Also, conditions like lower back pain can vary in their intensity from a twinge to debilitating agony, so, to answer the question, I would need to have some idea about how severe the condition is. Some critics point out that the theoretical validity of the measure is dependent on a series of dubious assumptions – for example, that living for two years with lower back pain merits twice the score of one year with the condition. More concerning from a charity perspective is the likelihood that different populations will evaluate conditions differently, so using QALYs elicited from, say, the UK population may not be appropriate when assessing the benefits of donations to poorer counties in Asia and Africa. Within populations, QALYs are usually assumed to be the same for everyone, but some would argue that a quality-adjusted life year for a child or even a potentially highly productive person may be more valuable.[52] Moreover, QALYs don't take into account secondary benefits. If a person's blindness is cured, they may be able to earn more money to support their family, thereby improving the quality of life of their children. And what about donating to animal charities? Does the quality of life of a dog or a horse or a chicken matter as much as that of a human? If it does, then we need some way of assessing QALYs for other creatures.

On the face of it, QALYs look as if they provide a rather flaky foundation for decisions about charitable donations. But I'd still like the money I give to have the maximum effect in improving people's lives. So what should I do? As with most numbers that are subjectively estimated, the answer is to treat QALYs as very rough approximations. In the case of charities, this roughness is probably good enough to provide useful guidance on the best places to donate, at least for charities that aim to improve the lives of humans. Because there are such huge differences in the cost-effectiveness of different charities, errors in the estimation of QALYs often make no difference to an assessment that a donation to one charity will bring more benefits (or QALYs) than a donation to another. For example, the Oxford-based philosopher Toby Ord, who founded the charity Giving What We Can, has provided evidence that the benefits resulting from a donation to one charity can be as much as 15,000 times greater than the benefits of donating to an alternative.[53] (In this case, benefits were measured by DALYs – disability-adjusted life years – a measure closely related to QALYs.) Even donations to charities that aim to tackle the same problem can vary hugely in their impact. For example, consider the aim of preventing and treating HIV and AIDS. In this case, funding education for high-risk groups, such as sex workers, can be 1400 times more cost-effective than funding surgical treatment for Kaposi's sarcoma, a rare type of cancer associated with an advanced AIDS infection that causes skin lesions and other symptoms.[54] That's a big enough difference to mean that QALYs can be of value in choosing between charities, despite their bluntness. In addition, it should be stressed that effective altruists don't just rely on QALYs or similar measures. They also apply criteria such as asking whether the target of a charity is a neglected area, and hence has room for more funding, and what the chances are that a donation will achieve what it aims to achieve. Websites associated with effective altruism, such as givewell.org and givingwhatwecan.org, also base their advice on additional factors, including a charity's track record and how transparent it is about its operations.

All this guidance is useful, but next time there's a major catastrophe I'll be torn. My heart will be urging me to donate to the disaster fund, but my head will be telling me to send that money to where the numbers suggest it should go: to help to mitigate those ongoing tragedies that seldom, if ever, make the headlines, but where even small donations can achieve so much more.

So can you trust subjective numbers?

As we saw with the measurement of pain, subjective numbers can be trusted when people produce consistent values under the same conditions *and* when the number is measuring what it is supposed to be measuring. Researchers call the first of these qualities reliability. When I mark students' work, I have to give the same mark to students who produce the same answer or my marking would not be reliable. But reliability on its own isn't enough. While I might have given the same mark to the two students, it may have been too high or too low given the quality of their answer. For my marks to be trustworthy, they also needed to meet the second quality, known as validity: they should reflect what they were intended to measure, namely the quality of the students' work.

We've seen that there are many reasons why subjective numbers can fail to be reliable and valid. The phrasing of questions and the order in which they appear, and questions which deal with ill-defined concepts that can be interpreted in different ways, can all influence our judgements. Then there's our inability to answer a question honestly or accurately. Many of us are more interested in pleasing and impressing an interviewer than giving valid responses. Often our assessments are influenced by recent events or ephemeral thoughts. Sometimes we genuinely just don't know the answer, but we think we do. At other times we might think that any old answer will do just to get the assessment over with.

Nevertheless, it's worth noting that there are some subjective estimates that people are quite good at. Ask people about the relative frequency with which different events occur and you are likely to get an accurate answer. One study asked people to estimate the relative number of restaurants in fast-food chains and their answers were closely aligned with the actual numbers.[55] People are also adept at estimating the frequency with which letters and words occur. Why is that? It might have something to do with our evolutionary past – for example, to survive it could have been helpful to estimate how frequently one encountered a prey animal when hunting in hot weather or how often people with a particular symptom went on to develop a disease.

Estimating group sizes accurately is another skill that we may have acquired from our distant ancestors, who would have needed to make decisions on whether or not to do battle over resources with other tribes, depending on their size. For example, German participants produced reasonably accurate estimates of the number of German cities that had populations in different size categories, such as 100,000 to 199,999 and 200,000 to 299,999.[56] Then there are the examples where instant intuitive predictions by people, based on very small amounts of information, have proved to be remarkably accurate. We'll look at these in the last chapter.

But what should we do when we need to make a decision based on subjective numbers and we suspect that they may be less than perfectly accurate? We saw with QALYs that some charities remain more cost-effective than others even if there are large errors in the estimation of the relevant QALYs. Decision analysts use a similar method called sensitivity analysis, which involves assessing how robust a decision is to errors in subjective numbers. For example, we might find that, even if a person's rating of their preference for a car on a 1 to 10 scale is 30 per cent out, they would still prefer that car to an alternative.

So, despite their deficiencies, subjective numbers can often be useful when we only need a 'rough and ready' idea of what might be the truth, but, because of their coarseness, we should also be suspicious of very exact numbers based on subjective responses. When newspapers tell us that people reckon they need to win exactly £22.3 million on the lottery to achieve perfect happiness,[57] that married couples are at their most blissful two years, eleven months and eight days after their wedding,[58] and that British people stay in a sulk for an average of seven minutes after being served a substandard cup of tea, it's probably wise to reach for a pinch of sodium chloride.[59]

8

It's Probably True: Soft Numbers Meet Hard Data

Subjective numbers in science: The 'Great Satan'

Ronald Aylmer Fisher (whom we met in Chapter 4) was, by all accounts, a complex character. A brilliant thinker – he made major contributions to statistics, the design of experiments, and genetics – and occasionally warm and charming, he was also beset by personal bitterness and animosity. Though he was himself passionate in the pursuit of scientific truth, Fisher's superstitious mother insisted on his birth in 1890 that his name should contain a 'y'; an earlier child, called Alan, had died at a very young age, but his older siblings, Geoffrey and Evelyn, had survived. Famously myopic – he was rejected for military service in 1914 – Fisher also suffered from an uncontrollable temper. At 27, he insisted on marrying his 17-year-old bride, Ruth Eileen, in secret and without her widowed mother's knowledge or approval. This obstinate trait in Fisher's character would later lead to him being involved in a number of acrimonious disputes with other leading figures in the developing field of statistics.

One of Fisher's battles related to the role of subjective assessments of the truth in science. There are, of course, many situations where we can't know the truth for certain. A doctor might diagnose that numbness in a patient's hands and feet means that they have type 2 diabetes, but she can't be sure. A TV interviewee might claim that 3 per cent of the British population are allergic to peanuts, but does he have a vested interest in exaggerating the figure? A chemical company might rebut reports by sceptical scientists who assert that a much-hyped new fertiliser will lead to no increase in the yield per acre of wheat. But, surely, the company has a vested interest in confronting the scientists? A politician might argue that the majority of voters favour reduced spending on overseas aid, but is he simply telling his constituents what they want to hear?

The way forward in these circumstances is to gather data and weigh up what this is telling us. This might involve carrying out a survey on a representative sample of the population. We might, for example, survey 1000 British people and establish whether they have a peanut allergy. Or we might apply the fertiliser to a sample of ten randomly selected field locations and compare the yield with that of fields that have not been fertilised. But, because our evidence is only based on a sample, it will seldom tell us for certain what the truth is.

In cases like this, Fisher suggested that we should try to get an assessment of what the truth might be by first formulating a hypothesis.[1] This usually involves tentatively assuming that a claim is true until the data suggests that it is implausible. So we would, for example, hypothesise that 3 per cent of Britons *do* suffer from a peanut allergy. We should then gather some relevant data. Having analysed it, Fisher suggested we should look at any discrepancy between our survey result and our hypothesis and ask: 'How probable is it that we would get a discrepancy at least as large as this if the hypothesis is true?'[2] Suppose our sample of 1000 people finds that only 1 per cent have a peanut allergy. One explanation for the discrepancy between the 3 per cent claim and our 1 per cent survey result is that, by pure chance, the people we selected in our sample had a lower incidence of the allergy than the general population. All estimates from samples have margins of error, so the claim may still be true; the difference might just be a typical error that occurs when we take samples. The other explanation, of course, is that the claim is false.

If we selected people for our sample randomly, we can use probability theory to calculate the chances that we would have a discrepancy from the true value at least as large as 2 per cent (i.e., 3 – 1 = 2 per cent). Suppose this probability is only 1 in 1000. We could now say to the pundit who made the claim: 'Look, if your claim is correct, then there is only one chance in a thousand that our sample estimate will be at least 2 per cent different from your claim – so we have strong reasons to doubt your claim. It's very unlikely we would have got this result in our survey if your claim is true.' (Recall that the one in thousand, or 0.001, is called a *p*-value – we met these earlier, in Chapter 4.) The key point here is that Fisher's method does *not* tell us the probability that the claim (or hypothesis) is true. It only tells us the probability of getting the survey result that we did *if* the claim is true, which is a different matter.

To take another example: the doctor's diagnosis. She decides to take a blood sample that will indicate whether her patient has type 2 diabetes. This suggests that they have the disease. But the test has only an 80 per cent chance of giving a correct indication, so all they know is: if they have diabetes, then there is an 80 per cent chance that the test will give the indication that it did. This is a rather tortuous statement. What they really want to know is simple: what are the chances that they have diabetes now that they've had the blood test?

There are many circumstances where we would prefer to know the probability directly that someone's claim or hypothesis is true, instead of some statement that tells us about the reliability of the evidence rather than the reliability of the hypothesis itself. What are the chances that the claim that the fertiliser is useless is true? What is the probability that a drug is harmless? What are the chances that as many as 3 per cent of British people are in danger from peanuts? But Fisher's method, despite being the basis of much scientific work, can't tell us these things. To answer the question that we want answered, we need to turn to a theorem developed by an eighteenth-century English clergyman, Thomas Bayes.

Little is known about Bayes, apart from the fact that he was born in 1702, practised as a Nonconformist minister in Tunbridge Wells in Kent, and was elected a Fellow of the Royal Society in 1741. However, his most important work, which was published in 1763, two years after his death, has been hugely influential. It shows how we should update our beliefs in the light of new information, such as information from a medical scan or a survey. For example, you look at the cloudy morning sky and estimate that the chance of it raining today is 60 per cent. You then watch the television weather forecast and are surprised to find that it predicts fine weather all day. However, you know that the TV forecast is not foolproof – on 10 per cent of days it gets the forecast wrong. Bayes's theorem shows that, having seen the forecast, you should revise your probability of rain down to just 14 per cent. (Bayes's theorem for this sort of problem is easy to apply; the notes show how this last figure was arrived at.[3])

Some historians believe that Bayes was trying to use probability theory to show that reported miracles were not fakes or illusions, as had been asserted by the contemporary philosopher David Hume. Others claim that he was using probabilities to try to prove the existence of God (more on this later). One theory is that it was Bayes's close friend, the 'quiet, unassuming' Welsh minister Richard Price, who first applied the theorem to these religious questions. It may even have been the case that Price discovered a half-complete version of Bayes's theorem when he went through Bayes's unpublished papers after his death and that it was he who should be credited with perfecting the theorem (indeed, a Richard Price Society exists that aims to recognise Price's contributions).[4] Whatever the truth, these days you will find Bayes's theorem being put to work in applications as diverse as spam filters, oil exploration, artificial intelligence, estimating the age of the universe, and forecasting the success of a movie. In 2003, the theorem was used to find a missing lobster fisherman, John Aldridge, who was lost off Long Island near New York.[5] But how does Bayes's theorem relate to Fisher's method? And why did the theorem rile Fisher to the point where condemning it became an obsession?

The answer is that the theorem allows us to use our judgement, experience and expertise to put an initial probability on the possibility that a hypothesis is true (this is called a prior probability and is a reflection of our degree of belief that a hypothesis is true). The theorem then tells us how to revise our probability when we get the results from our sample or survey. Unlike Fisher's method, we end up with a probability for our hypothesis – one based on a combination of our subjective judgement and the hard data from our sample or other source.

In the case of the patient with numbness in their hands and feet, suppose that the doctor has lots of experience and used her judgement to estimate, before receiving the result of the blood test, that there was a 70 per cent chance that the patient had diabetes. The result from the blood test then also indicates that the patient has diabetes, and this has an 80 per cent chance of being accurate. When we combine these two pieces of information, Bayes's theorem tells us that the probability that the patient has diabetes is over 90 per cent.

And that's what irked Fisher – the idea that people were allowed to use their subjective judgement to help to determine the plausibility of a hypothesis.[6] Science, in his view, needed to be objective. The blood test result is hard evidence; the doctor's prior estimate is a 'soft' judgement and therefore, according to Fisher, should be inadmissible in scientific research. Since the 1920s most scientists have echoed Fisher's view. As science writer Robert Matthews wrote, 'to most scientists, subjectivity is the Great Satan'.[7]

The myth of objectivity

Despite scientists' claims that they abhor subjectivity, the idea of the white-coated researcher coolly working away in their laboratory to test hypotheses in a dogged and disinterested pursuit of the truth, come what may, is a myth. Science is unavoidably subjective, even though this subjectivity is rarely made explicit.[8] As humans, scientists have preconceptions, beliefs, careers to pursue, reputations to consider, and egos. In some cases this can even lead to them engaging in outright fraudulent behaviour – fiddling their data or dishonestly reporting their analyses.

It was reported in 2012 that since 1975 there had been a tenfold increase in the number of biomedical and life sciences papers retracted from journals because of scientific misconduct (over 1300 fraudulent papers were discovered in the study).[9] However, even honest scientists can unwittingly allow their subjective beliefs to influence their findings. Designing and running experiments involves many decisions, some of which are judgement calls.[10] Which equipment will be used? How large a sample should be taken and who or what should be sampled? Which extraneous variables should be controlled? For example, if we want to compare the reaction times of old and young people, should we control for their gender, their intelligence, or the time of day when they participated in the experiment?

Perhaps most open to judgement is the problem of how to treat strange or unusual observations that occur during an experiment. In one experiment I helped to conduct, we wanted to find out how good people were at predicting the increase in demand for products that would occur when a supermarket ran a sales promotion. We supplied the participants with details about the promotion, together with past data that showed the sales uplifts that had been associated with earlier promotion campaigns. In every past case the campaign had been successful in increasing sales. Despite this, a small number of participants in the experiment predicted that the promotion would decrease sales or make no difference (in a similar experiment some people predicted spectacular increases in sales that were just not plausible based on the available information). There seemed to be no rationale for these odd results. Had the participants misunderstood the task? Were they, for some obscure reason, trying to wreck the experiment? Had they mistakenly hit the wrong buttons on the keyboard? We decided to exclude these people from the results, but in the interests of transparency, we reported what we had done.

Nevertheless, the dangers are clear. Researchers are likely to read into data what they expect to see. Depending on their beliefs and their perspectives, different people will see different messages in the same data. These messages may come across as objective findings. One person's random blip in a graph might be another's evidence that a theory needs modifying. Sometimes the unusual observations that we dismiss might be the most significant. Think of poor Kermit A. Tyler, a first lieutenant on temporary duty at a radar information centre on the Hawaiian island of Oahu on a quiet and seemingly uneventful Sunday morning in 1941.[11] Two radar operators reported seeing an abnormally large blip on their radar screen which suggested that a large number of aircraft were rapidly approaching from about 130 miles away.

'Don't worry about it,' Tyler reassured the operators, thinking it must be a flight of American bombers that he knew were on their way from the US mainland. Those few words would dog Tyler for the rest of his life. The radar had, in fact, detected more than 180 Japanese aircraft that were closing in for a deadly surprise attack on Pearl Harbor, an event that would thrust America into the Second World War. Although subsequent official inquiries exonerated Tyler of any blame – he was untrained for the role he was playing on that fateful morning and the arrival of the US bombers was an entirely plausible explanation for the blip – he still received occasional furious letters condemning him for his inaction.

Like quantum phenomena, whose characteristics are not determined until they've been observed, data has no character and no message until it's been interpreted by a human. And in the supposedly objective arena of science, that interpretation will inevitably be subjective.

The case for Bayes

At a first glance, explicitly allowing people to put their subjective probabilities into scientific analysis might seem worrying. Couldn't the results be manipulated to confirm their prejudices? We also saw in Chapter 7 that subjective numbers can be inconsistent, biased and unreliable. This is particularly the case when people are asked to estimate probabilities in their heads. The psychologists Amos Tversky and Daniel Kahneman became famous for their work that identified and explained the causes of many of these biases.[12] They showed that our estimates of probabilities can be over-influenced by recent or easily recalled events, by anecdotal information and stereotypes, and by initial numbers that somehow get implanted in our heads.

However, we have seen that subjectivity and its attendant biases are unavoidable in science, so surely it's much better to bring this into the open in a formal and documented way. Bayes's theorem provides a structure for this. It requires people to be explicit about their subjective beliefs prior to gathering more information. And if we ask scientists to articulate and record the rationale for these initial beliefs (based on earlier research results, for example), then we will be in a far better position to judge the reliability of their findings. Hiding subjectivity under an informal cloak of objectivity does not achieve this.

Bayes's theorem has other attractions. It neatly balances how confident we are in our subjective judgement with how reliable the new information is. Highly reliable new information will usually have a relatively large influence on the revised probability (this is elegantly referred to as the posterior probability). Similarly, if we are not confident about our prior probability and say that our hypothesis has a mere 50 per cent chance of being correct, Bayes's theorem will base its revised probability solely on how reliable the new information is. If, instead, we are prepared to stick our neck out and say that our hypothesis has a 98 per cent chance of being true, then our prior probability will have a huge influence on the revised probability. When we go to the extreme and say we are 100 per cent certain that a hypothesis is true, then any new evidence, however reliable, will be ignored and the revised probability will be 100 per cent. That would not make much sense. If I say I am 100 per cent sure that flying saucers have landed at Buckingham

Palace, Bayes's theorem would not revise this probability, even when highly reliable news sources tell me that it hasn't happened. However, as long as I don't go to this extreme, as more and more hard data arrives it has a bigger and bigger influence, relative to our initial estimate, on the posterior probability of a hypothesis being true, which makes sense. This also means that if researchers disagree about their prior probabilities, their posterior probabilities will tend to converge as the data accumulates.

If we don't take account of the initial plausibility of a hypothesis, we can expose ourselves to seemingly astonishing findings that turn out to be based on mere flukes in the data. I might have a madcap theory that people who eat porridge in the morning develop a higher IQ. When I test the IQ of 20 regular porridge eaters and 20 people who can't stand porridge, I find that the porridge devotees, on average, have an IQ that is five points higher than the porridge haters. Better still, I do the calculations and find that the probability of finding a difference at least this large, if porridge really has no effect on IQ, is only 3 per cent. Happily concluding that this implies porridge does have brain-enhancing properties, I rush to get my crucial discovery published in the literature and then wait by the phone for the media attention this will surely bring me, not to mention the accolades from the porridge industry. In reality, it's highly likely that my theory is nonsense and a dose of plausibility fed into the calculations via Bayes's method would soon have helped to dispatch it to the scientific scrap heap. The higher IQ of the porridge eaters I studied was almost certainly just a fluke.[13]

The worry is that, of the millions of science and social science papers published annually, a large number are bound to be based on such flukes.[14] Yet many of these papers will be influential and their findings may enter the public consciousness as hard facts, difficult to dislodge. The widespread presence of freak results is one reason why scientists searching for new cancer drugs have often failed to reproduce other scientists' landmark discoveries in their own laboratories. Amgen, a large American biopharmaceutical company, found that the results of 47 out of 53 studies could not be replicated – a worrying finding for cancer research.[15]

In psychology there is now said to be a replication crisis. Most people think this began around 2011 when the psychologist Daryl Bem published a paper in a highly reputable journal providing evidence that people have powers of extrasensory perception (ESP) that allow them to see into the future.[16] One of his experiments had undergraduates at Cornell University sitting in front of a computer monitor which displayed an image of a pair of curtains. Behind one of the curtains was a pornographic image, selected because Bem believed that, if we did have sixth sense, it would have ancient origins and hence would be responsive to our most basic drives. The undergraduates' task was to decide which of the curtains concealed the image. Crucially, this was randomly determined only *after* the students had made their choice, so if they got it right, they had correctly predicted a future event. The predictions were correct just over 53 per cent of the time. This doesn't sound impressive, but it is far better than the 50 per cent success rate that they would have expected to have achieved by guessing. Bem calculated that there was only 1 chance in 100 that they could have done as well as this based on guesswork. Altogether, his paper reported the results of nine experiments involving a total of more than 1000 participants and eight of these yielded results that were unlikely to be explained by guesswork. The findings seemed unbelievable, but Bem's research methods were carried out honestly and transparently and followed standard accepted research procedures. Many psychology researchers were alarmed that the methods they had employed for years could make the impossible appear to be a reality. A major and carefully implemented attempt to replicate Bem's results in 2017 found no evidence at all of the phenomenon.[17] Not that this was much comfort: a 2015 investigation found that only about 40 out of 100 studies published in top journals had results that could be replicated.[18] The Nobel laureate Daniel Kahneman, himself a psychologist, warned of 'a train wreck looming'.

A good rule of thumb is that, the more newsworthy a finding is, the less likely it is to be successfully replicated. For example, psychologists at the University of Kentucky conducted an experiment which found that getting people to think more rationally made them less likely to state that they were religious.[19] The participants were Canadian undergraduates who were asked to rate their belief in God on a 0 to 100 scale. Before providing their rating, 26 of the undergraduates were asked to look at a picture of Rodin's statue *The Thinker*, which depicts a bronze figure in a reflective thinking pose, before making their rating. Another 31 were shown, instead, a picture of an ancient Greek statue, Myron's *Discobolus*, which depicts an athlete throwing a discus. Those who saw the picture of *The Thinker* gave a lower rating for their belief in God – and there was only a 3 per cent probability that the difference could be explained by chance. The researchers concluded that seeing the picture of *The Thinker* had primed the participants to think rationally, and this, in turn, had led them to express more religious disbelief.

It seems incredible that merely looking at a picture of a sculpture could have such a profound effect on people's stated beliefs, but the paper was published in one of the world's most prestigious journals, *Science*, and has since been cited in other papers nearly 400 times. It turned out that when other researchers tried to reproduce this sensational finding, they failed, suggesting that it was a statistical fluke, especially given the small number of participants involved.[20] The same researchers also failed to replicate another study published in *Science* that reported that washing your hands stops you worrying about past decisions and lessens your need to justify a recent choice to yourself.[21] The theory was that washing your hands removes 'psychological traces of the past' so you stop mithering about it.

What to make of all this? One lesson is that, until scientific findings have been replicated several times, we should treat them as provisional. So, defer judgement on recent findings that walking backwards improves your short-term memory,[22] or that your first name influences how competent or warm people think you are.[23] But trying to replicate other researchers' work is a relatively unglamorous task – you won't be the person credited with the world-changing discovery if it is validated. Academic journals also prefer hot new exciting findings, as they're more likely to attract citations in papers by other researchers and maintain a journal's ranking. Besides, the argument goes, once research has been published, surely it should be trusted; after all, it went through a lengthy process of peer review by experts in the field before being accepted. And the researchers applied Fisher's method, like most other researchers, so the findings must be solid. Why bother to replicate them?[24]

Even when it does take place, replication may not be enough to ensure that scientific findings are robust or even to tell us they are wrong. Scientists from a similar cultural background to the original investigators might bring the same unconscious biases to the problem, with the result that false findings continue to appear true. Or factors that the researchers haven't thought of may mean that findings fail to be replicated – for example, seemingly trivial things, such as the temperature of the room, may lead to different findings.[25]

Adding Bayes's theorem to the scientific toolkit, alongside replication, would increase the protection we have against the astounding finding that turns out to be baseless, or the see-saw reports in the media that tell you one day that something is good for you and a year or so later that it is very bad and should be avoided. The Austrian-born philosopher Karl Popper once wrote: 'Science must begin with myths, and with the criticism of myths.' Indeed! But the creation of new myths is surely something that modern science should take reasonable steps to avoid.

Postscript: Two unusual applications of Bayes's theorem

The existence of God

In April 2004, newspapers carried headlines such as: 'Is there a God

– it's 2/1 on' and 'Is there a God? Maths suggests it all adds up'. They were referring to a new book by Stephen Unwin, a former researcher in quantum gravity who had a job calculating the probability of nuclear accidents for the US Department of Energy and who later worked as a risk consultant. Among the positive reviews of the book was a warning from Rob Grant, the comedy writer who helped create the *Red Dwarf* television series, that: 'This book is very bad news for anyone planning a career in Evil.' Unwin called his book, which was praised for its breezy style and self-deprecating humour, *The Probability of God: A Simple Calculation that Proves the Ultimate Truth*.[26] The simple calculation was based on Bayes's theorem.

Starting with what he called a position of 'absolute ignorance', Unwin gave God a 50 per cent chance of existing – this was his prior probability. He then identified a number of factors that might provide evidence for or against the existence of God. For example, humans have a sense of goodness, people can be wicked, miracles such as the resurrection occur, and nature can do evil things through phenomena like earthquakes, floods and forest fires. For each of these bits of evidence he estimated how much more or less likely they were to exist if there was a God, compared to a universe where there was no God. For example, he thought that human goodness was ten times more likely in a universe where God existed; natural evil was, he estimated, ten times more likely to exist in a universe where there was no God. He then used Bayes's theorem to revise his prior probability in the light of this evidence. The result: there is a 67 per cent chance that there is a God.

Unwin admitted: 'This number has a subjective element since it reflects *my* assessment of the evidence.' In fact, the entire calculation is subjective, including the choice of factors that Unwin used as evidence for or against a God. Even if we agreed on his choice of evidence, different people would, of course, put different numbers on it. And even the choice of a 50 per cent prior probability, which appears to be neutral, is a subjective value. Most of us could justify starting off from different positions before we turn to the evidence

Unwin has used. While this is an interesting example of structured and documented thinking – and there's nothing wrong with that per se – the much publicised 67 per cent suffers from a lack of concrete foundations. Soft numbers have met soft numbers, and there is not a hard number to be seen.

Guilty or not guilty?

Imagine finding yourself in court accused of a murder you didn't commit. The only evidence against you comes from a reliable witness who described the murderer as a person who had the same hair colour, height and gender as you. Like you, they also wore spectacles. A statistician from a local college has calculated that only 1 in 80 people would match this description. The lawyer for the prosecution, with a hint of triumph in his voice, argues that this means that there is only a 1 in 80 chance that you are innocent, so there are 79 chances out of 80 that you are guilty. Not surprisingly, the jury find you guilty.

What went wrong? The answer is that the jury were deceived by what is known as the prosecutor's fallacy. The statistician had calculated the probability of the murderer matching your description given that you are innocent. However, despite what the prosecutor said, this is not the same as *the probability that you are innocent* given that you match the murder's description – and this probability, the reverse of the previous one, is the probability the jury needed to see.

Fortunately, Bayes's theorem can come to the rescue. If the police estimate that around 1000 people were in the area at the time of the murder, including you, then the prior probability that you were the murderer is only 1 in 1000. Updating this in the light of the witness's evidence yields a revised probability of just over 92 per cent that you are innocent – hopefully enough to convince the judges when you appeal![27] The logic is simple. If there were 1000 people around when the murder took place and 1 in 80 match the description of the murderer, then there would typically be about 13 people (1/80 × 1000) around who fitted the description provided by the witness. Of these, 12 are innocent, so the probability that you are one of the innocent people is 12 chances out of 13, or 92 per cent.

The prosecutor's fallacy is far from being a mere academic concern. It was evident, alongside other abuses of statistics, in the famous Collins trial in California in the 1960s, when a married couple were wrongly found guilty of robbing an elderly woman, based on descriptions of the assailants. The jury was swayed by the prosecution's erroneous claim that the probability that they were innocent was only 1 in 12 million. For similar reasons, a British solicitor, Sally Clark, was wrongly convicted in 1999 of the murder of her two infant sons. In her trial it was erroneously claimed that the probability that both deaths resulted from sudden infant death syndrome, rather than murder, was only 1 in 73 million, implying that this was the vanishingly small likelihood that Clark was innocent. Both cases were later overturned, but Sally Clark spent three years in prison. She never recovered from her experience and four years later she died of acute alcohol intoxication, though there was no evidence of suicide. She was 42.

Despite these miscarriages of justice, in 2011 a British judge banned the use of Bayes's theorem in court, complaining that the underlying statistics lacked 'firmness'.[28] This caused alarm among statisticians, but in 2013 judges in the Court of Appeal in England similarly rejected arguments based on Bayes's theorem. Their reason: 'You cannot properly say that there is a 25 per cent chance that something has happened ... Either it has or it has not.'[29]

Thomas Bayes (or should that be Richard Price?) would probably have disagreed.

9

To Hell with the Numbers

Catching a killer

On the edge of Greater Manchester, where the Pennine hills to the east rim the horizon like low clouds, lies Hyde, a town of red-roofed suburban houses and ivory-white apartment blocks. Dr Harold Shipman, grey-bearded and bespectacled, with a brisk, imperious, 'no nonsense' air about him, was popular with his patients in the town, while his colleagues regarded him as 'practising excellent medicine'. Few knew that he had a conviction in the 1970s for forging drug prescriptions for his own use; he seemed the essence of respectability. He even appeared in a national television documentary, advising how people suffering from mental illness should be treated within the community. Yet, for a quarter of a century from the mid-1970s, Shipman had been injecting many of his patients – mainly elderly women – with lethal doses of diamorphine and then falsifying their medical records to indicate they had died of natural causes.

Shipman was only caught because of the persistence of the daughter of one of his victims. Angela Woodruff was a solicitor who looked after her mother's affairs. Looking through her mother's legal documents, she was shocked to find a will that made Shipman the main beneficiary of the estate. Certain that the will was a forgery, and suspicious that the doctor had murdered her mother to profit from her death, she contacted the police. When her mother's body was exhumed, a post-mortem revealed that the cause of death was a morphine overdose. It was established that this had been administered around a time when Shipman visited her. The police searched the doctor's house and found an old type-writer with a broken key that he had used to forge the will. In an atmosphere of public incredulity and horror, the police were soon able to link more and more deaths to Shipman. In January 2000, after a trial lasting nearly three months, he was convicted on one count of forgery and fifteen counts of murder. Four years later, prison officers found him hanged in his cell at Wakefield jail.

Shipman may have murdered at least 260 people, though the exact figure will probably never be known. Despite this huge number, it's easy to see why he got away with his crimes for so long. We tend to see what we expect to see, and no one at the time would have expected a well-regarded family doctor to be a mass murderer. Shipman was also cunning in covering his tracks and convincing when challenged.

But could statistics have prevented the deaths? Evidence was presented at a public inquiry suggesting that it might have done if the data had been available. Surprisingly, the statistical methods that could have trapped Shipman were developed in the Second World War to control the quality of armaments production. They involved the use of charts to display over time whether the number of defective armaments in a batch exceeded the number that would normally be expected, which would indicate a decline in production standards. In medicine these methods can be adapted to show the extent to which the number of deaths associated with a doctor exceeds the number that would be expected for a practice of a similar type; this is called excess mortality. Excess mortality is plotted cumulatively against time, and if the line crosses an upper threshold, it signals that there *might* be a problem with that doctor. The threshold is set so that it balances the risks of a false alarm – when an entirely innocent doctor would be investigated – with the risk of not detecting a doctor like Shipman. Of course, some doctors may have higher than average levels of excess mortality by chance, so the chart is designed to signal when the excess mortality is very unlikely to be attributable to chance factors.[1] The statisticians David Spiegelhalter and Nicky Best estimated that, in theory, a chart like this could have stopped Shipman as early as 1985, well over a decade before he was arrested, during which time he committed possibly 200 of his murders.[2] Today, largely as a result of the Shipman case and other tragedies, monitoring methods like this have been put in place in healthcare systems in many countries to track the performance of general practitioners, surgeons and hospitals.[3]

Much of this book has demonstrated the ways in which numbers can be abused or overemphasised so that we are presented with a distorted view of reality. But there's another danger. This occurs when accurate, honest numbers do have something important to tell us, but either we fail to record them or, if they are available, we discount or ignore them. In areas ranging from political decision making to fraud detection and from planning large projects to medical diagnosis, we can disregard these numbers at our peril.

In the next chapter we'll see how closing our eyes to accurate numbers can make our world seem more dangerous than it is, even making us change our behaviour, at some cost, to avoid risks that are vanishingly small. In this chapter, we'll look at how we often fail to benefit from the valuable information contained in numbers. As we'll see, when we have the opportunity to update our beliefs as new information becomes available, we often behave very differently from the ideal represented by Bayes's theorem, which we met in Chapter 8. In effect we often set our prior probabilities at one or zero, indicating that we believe something is certain or impossible. In this case, no new information, however reliable, can make us change our minds.

We don't want to believe this, so we won't

Shipman was able to go on murdering people year after year because people assumed that a doctor would never deliberately harm his patients. So no one collected or analysed the statistics that could have exposed him as a mass murderer. But disbelief can also cause us to fail to act when reliable statistics are available, and even when they are writ large, shouting at us to act or change our ways. Once we have a settled view of the world, most of us tend to resist any challenges to it. We might try to ignore the numbers we don't want to hear or do all we can to contest them, distort them, or dismiss their relevance.[4] Evidence of global warming and its causes are rejected by many people. Others persist in their beliefs that vaccines cause autism or cell phones cause brain cancer, despite reliable evidence that they don't. But why are so many of us so reluctant to change our minds when the facts tell us that we should?

One theory proposed by the cognitive scientists Hugo Mercier and Dan Sperber, of the French National Centre for Scientific Research and the Central European University in Budapest respectively, is that our ability to reason didn't evolve so that we could discover truths. Instead, as social animals, reasoning developed so we could justify our actions and decisions to others, thereby improving communication and cooperation.[5] Being able to defend our position in arguments also served to increase our prestige and status within our social group. Hence there was an incentive to overweigh evidence and choose lines of argument that favoured our actions while simultaneously downplaying inconvenient facts.

In addition, the philosopher Julian Baggini points out that most of us have webs of mutually supporting beliefs. A belief in individual responsibility might support a belief in low taxation, the absence of a welfare state, long sentences for crimes, and small government. Or a belief in nurture rather than nature might, in turn, support beliefs in high educational expenditure and income redistribution. If one of these beliefs is questioned, a person's whole network of beliefs and values might unravel. As Baggini argues: 'Challenge someone's truth and often you challenge their whole world.'[6]

One consequence of all this is the well-known phenomenon of confirmation bias, where people selectively seek out facts to confirm their beliefs, while neglecting the possibility of finding information that may prove these beliefs to be wrong. There is even some evidence that we experience a dopamine-fuelled rush of pleasure when we find information suggesting we are right, even if, in reality, we are wrong.[7]

The opposite is true when we are confronted with disconfirming information. Cognitive dissonance refers to the uncomfortable feeling we have when we hold inconsistent or contradictory beliefs at the same time. Avoiding or ignoring the unwelcome new information is one way of alleviating the discomfort. It's easier than reformulating our world view. We seem particularly sensitive to arguments opposing our political beliefs. One brain scan study – albeit involving a small number of people – found that when our political ideologies are challenged, parts of our brain associated with self-identity and negative emotions are activated, and we interpret these challenges as personal insults.[8]

Perhaps this can explain why, when we encounter arguments and evidence that contradict our beliefs, they can become deeper and even more entrenched. Psychologists call this the 'backfire effect'. In an experiment conducted in 2005 by Brendan Nyhan, professor of public policy at the University of Michigan, and Jason Reifler, now professor of political science at the University of Exeter, people were presented with mock newspaper articles that included a misleading claim from a politician.[9] One article stated that Iraq had weapons of mass destruction (WMD) immediately before the US-led invasion in 2003. This was a belief that persisted with some Americans even after the war was long over. A little later, some of the participants were shown a newspaper article that corrected the previous report by indicating that there was no evidence of stockpiles of such weapons or an active programme of WMD production. The results of the experiment indicated that those whose beliefs were consistent with the original misleading article were actually more likely to believe that Iraq possessed WMD after they had read the correction.

One strategy people employ when confronted with numbers that run counter to their beliefs is to reframe the issue as one where the facts are irrelevant. In another experiment people opposed to same-sex marriage were given a mock-up of statistics indicating that children raised by same-sex parents had the same outcomes (e.g., in terms of their career success, intelligence and criminality) as those of opposite-sex parents. They replied by arguing that this was a moral matter, so the data was not pertinent. Those in favour of same-sex marriage used the same argument when presented with data purporting to show that this led to worse outcomes for children.[10]

A resistance to reliable information can persist even in the most intelligent people, including those who are trained to think scientifically. It's claimed that there is a phenomenon called 'Nobel disease' that afflicts former winners of the world's most prestigious science awards. There are numerous examples of people who, after reaching the pinnacle of scientific discovery, spent the later part of their careers believing in bizarre things, despite the weight of contrary evidence. Charles Richet, who carried out innovative work in immunology that earned him the Nobel Prize in Physiology or Medicine in 1913, spent much of his life believing in the paranormal; he invented the word 'ectoplasm'. More recently, Kary Mullis, who won the 1993 Nobel Prize in Chemistry for his hugely important work in biochemistry, denies that the HIV virus causes AIDS, believes that climate change is not caused by human activity, and even claimed to have encountered a 'glowing raccoon' one night in the woods near his Californian cabin, and which addressed him with the words: 'Good evening, doctor.'[11]

Then there is Wolfgang Pauli (Physics, 1945) who believed that his mere proximity to equipment could cause it to fail (the so-called Pauli effect). As a result, some colleagues banned him from their laboratories. On one occasion, as a prank, some fellow physicists arranged for a chandelier to crash to the floor when Pauli entered the room. It failed to do so, ironically seeming to provide more evidence for the Pauli effect. Other beliefs espoused by Nobel laureates include homeopathy, creationism, mysticism, psychic powers, and the effectiveness of high doses of vitamin C as a treatment for the common cold and cancer.

We know best

The barriers we erect to protect ourselves from accurate numbers can also leave us prone to rejecting good advice, or at least not taking it fully into account. When receiving numeric advice on topics ranging from the potential future sales of a product to the answers to general knowledge questions, people tend to give undue credence to their own opinions. This tendency has been called 'egocentric advice discounting'.[12] It can occur because we have direct access to our own reasons for taking a particular point of view, whereas the rationale of an advisor is likely to be more obscure.[13] Also, our egos may simply be geared to telling us that our own opinions are superior to those of other people.[14] Although we may grudgingly move some way in the recommended direction – for example, we may adjust our forecast of a share price towards that suggested by an advisor – this is usually insufficient when the advice is highly accurate.

We are also generally too ready to dump an advisor if their previous piece of advice proved to be wrong, even if they have an accurate long-term track record.[15] It's akin to sacking the manager of a successful football team after one bad result. Advisors can be summarily discarded even if the recent inaccuracy results from random factors that they could not reasonably have foreseen. One reason for this tendency to 'fire from the hip' is that bad advice is more striking than accurate advice; it surprises us because we usually expect an advisor to be competent. It can be a long struggle for an advisor to establish a reputation, but one or two recent cases of inaccurate advice can prove fatal.

These days we can also receive advice from computer algorithms and there is evidence that this can be treated with even less credibility than that from a human advisor, irrespective of its reliability. With colleagues in Turkey and Scotland, I conducted an experiment where one group of people were given stock market forecasts provided by a financial expert (this was the truth), while another group were given the *same* forecasts but were told that they were produced by a computer. The participants exhibited less faith in the accuracy of the forecasts they thought came from a computer rather than a human expert.[16] In another experiment, conducted by researchers in Australia and this time involving sales forecasting, people stuck to forecasts based on their own judgement despite being given accurate messages like: 'Please be aware that you are 18.1 per cent LESS ACCURATE than the [computer-based] statistical forecast provided to you.'[17]

These two experiments demonstrated examples of what is termed 'algorithm aversion' and it appears to occur because we are less tolerant of computers when they get things wrong than we are of humans.[18] Algorithms do, of course, occasionally make mistakes. Imagine that your normally reliable satnav advises you to change your route to work one day because of traffic congestion five miles ahead. You follow the advice, but as a result you get to work half an hour late only to hear from a colleague that there was a much quicker route. Compare this with a situation where it was *you*, not the satnav, who worked out the inferior alternative route. Research suggests that you will be much harder on the satnav. Lots of us are therefore too sensitive to the occasional rare error we experience with computer algorithms and rely on our own judgement instead. But in areas where the advice from algorithms is generally much more reliable, our aversion comes at a cost.

Just tell us the story

Sometimes numbers, including those produced by algorithms, have to compete with stories that paint an entirely different picture of the world – and often the story wins. Once I was observing a sales forecasting meeting in a company and the graph of past sales of a key product had been generally flat for months. But the most recent month had seen a sudden boost in the product's sales.

The sales forecasting algorithm used by the company dismissed the sales increase as a mere random twitch of the sort that can be seen on most sales graphs. It forecast continued flat sales. But the managers in the meeting were not impressed.

'Our new MBA-educated sales manager is already having an effect,' argued the most senior person at the meeting, rubbing his hands with satisfaction. The sales manager himself was absent from the meeting, so his blushes were spared.

'He impressed me with his energy and ideas at the interview,' another manager chipped in. 'A great appointment!'

'I agree – and he's already got a good relationship with our customers,' someone else added.

Gradually a story was developing to explain the upward jerk in the graph. Nobody pointed out that you shouldn't read too much into a single month's sales. To the managers, a sales paradise had already arrived and the conservative forecast from the algorithm was easily dismissed.

Stories are a natural medium through which we seek to make sense of the world and storytelling behaviour may have emerged between 30,000 and 100,000 years ago.[19] They explain motives, provide context, and relate to flesh-and-blood individuals rather than abstract averages of people. As such, they are more likely than bald statistics to engage our attention. Moreover, as the mathematician John Allen Paulos has pointed out: with stories we tend to suspend disbelief; with numbers we suspend belief.[20] And stories are easier to remember.

Ironically, the problem with stories is our sheer inventiveness as human beings. When confronted with a series of facts, we are brilliant at inventing explanations to link them together, just like the managers at the forecasting meeting. In fact, we hate finding ourselves in a vacuum where there is an absence of explanation. The Latin poet Virgil famously wrote: *Felix qui potuit rerum cognoscere causas,* which roughly translates as 'Fortunate is he who is able to know the causes of things'. The danger is that the invented story may too easily satisfy us so that it becomes the sole explanation for what we have heard or observed.[21] This is what the writer and scholar Nassim Nicholas Taleb refers to as the narrative fallacy. It leads to the illusion that we have a complete understanding of what underlies the facts when in reality we may be wide of the mark. As Taleb puts it: 'Statistics are invisible; anecdotes are salient.'[22]

Statistical victims

In October 1987 the world's attention was focused on the backyard of a house in Midland, Texas. CNN was broadcasting round-the-clock coverage. The president of the USA, Ronald Reagan, was following the events closely. The cause of this excitement was that an 18-month-old girl, Jessica McClure, had fallen down a dry well behind her aunt's house while her mother was distracted by a phone call. Encased in very hard rock, 22 feet below ground, in a space only 8 inches wide, Jessica's fate looked ominous: the first frantic attempts to rescue her proved futile. As efforts continued, people from around the globe sent flowers, gifts and monetary donations totalling around $800,000 to Jessica's family.

Eventually, 56 hours after her fall, Jessica, now referred to as 'Everybody's Baby', was carried alive from the well. A painting, called *A Triumph of the Human Spirit* by Jan Johnson Sheets, shows her being held aloft among her celebrating rescuers; in the background, sunlight blasts through the clouds.

Despite her rescue, Jessica needed extensive medical treatment and suffered some permanent injuries, but she grew up to have a family of her own. At the age of 25 she gained access to the trust fund that had been set up to hold the many donations she had received. Sadly, the fund was wiped out in the 2008 stock market crash, though by then Jessica had been able to buy a house.[23]

Here we have a heartening story of human perseverance, empathy and generosity. But as the anti-poverty campaigner Peter Singer pointed out, during those two and a half days, while the world waited and hoped for Jessica's rescue, an estimated 67,500 children died from poverty-related causes that could have been avoided.[24] Researchers have repeatedly found that people are much more likely to donate money to help a single person who is in trouble when they are identified, possibly with a name, age and picture, than multiple anonymous people who are facing the same plight.[25]

Dry abstract statistics don't arouse our emotions, but a story about an individual who needs our help is more likely to urge us into action. When Winston Churchill met Stalin at the Tehran Conference in 1943, in the midst of war, he allegedly objected to the early opening of a second front in France. Landing troops in Western Europe at that time, he believed, would result in heavy allied casualties. The Soviet tyrant is said to have replied: 'When one man dies it's a tragedy. When thousands die it's statistics.' Even Mother Theresa once said: 'If I look at the mass, I will never act. If I look at the one, I will.'

The psychologist Paul Slovic of the University of Oregon has investigated why we are so emotionally swayed by individual accounts of people who need help and so resistant to statistics about the numbers of victims when large disasters occur.[26] He sees two psychological processes in action. First, it is known that when a stimulus such as a light source gets stronger we detect the early increases much more than the later ones. Darkness to dimness is more noticeable than the same increase in the illumination of a light that is already bright. The same applies to our perceptions of heaviness, loudness, and even the value of money: we become less sensitive to increases as the quantity gets bigger. In the same way, as the number of victims in a disaster increases, each extra person evokes in us a smaller and smaller response. Slovic calls this phenomenon psychophysical numbing. It means that saving one life rather than none will seem more important to us than saving 99 lives rather than 98.

We also tend to pay more attention to the *percentage* of lives saved in a disaster than the raw number. This can lead to some curious results. For example, in an experiment conducted by Slovic and his colleagues, college students were randomly divided into two groups.[27] Participants in one group were asked to rate their support, on a scale from 0 to 20, for an air safety measure that would save 150 lives that were at risk. The second group were asked to indicate their support for a measure that would save 98 per cent of 150 lives that were at risk (that's 147 lives). The second group indicated stronger support, despite fewer lives being saved. Saving 150 lives is an abstract phenomenon that is difficult to grasp and evaluate, but saving 98 per cent of lives sounds good because it is close to the maximum that could be achieved. Even a safety measure that would save 85 per cent of 150 lives (that is, about 128 lives) received stronger support than one that would save 150.

But according to Slovic, there is more to the problem than this. Images and stories have a much more powerful emotional effect on us than mere numbers. They allow us to identify with the victim and feel for them. The resulting sensations – empathy, sympathy, compassion, sadness, pity and distress – are powerful motivators that urge us to help others.

In 2019, a caravan of over 6000 refugees from several Central American countries was massed on the US–Mexico border in hope of sanctuary. In response, Donald Trump sent troops to protect the country from what he characterised as an imminent 'invasion' of undesirables including 'criminals and unknown Middle Easterners'. A sad story to be sure, but it soon lost my attention when the TV news moved on to the next report. But then a *Time* magazine article told the story of Cándidón Calderón, who had fled Guatemala with his wife and three children, aged 9, 11 and 12. Gang members had threatened to kill Calderón's children if he didn't pay around $1200, which was nearly half a year's income from the juice stall the family ran. I could not get the horror of their plight out of my mind.[28]

Art can also change our perceptions of the world when the numbers seem cold and abstract. On 1 July 1916, the first day of the horrific Battle of the Somme during the First World War, 19,240 Commonwealth soldiers were killed. It is almost impossible to conceive of such a number, but the Somerset artist Rob

Heard began to wonder what that number really looked like. He wanted to 'physicalise' the number and started to create an effigy – a twelve-inch shrouded figure – for each person who had died. Every effigy is bent into a different shape, 'some with legs folded and others outstretched – each one as unique as the individual men they represent'.[29] The resulting artwork, with the figures neatly laid side by side in row after row in a large open space, is hugely effective in conveying the extent of the loss of life on that terrible day. Heard has since extended his artwork so that it now represents the 72,396 British soldiers and South African infantrymen whose bodies were never recovered from the battlefields of the Somme.

One of the most evocative images is that of a human face, but, as Slovic points out, even the face of an animal in trouble can unleash a tidal wave of compassion or protest. During the devastating foot and mouth outbreak in Britain in 2001, millions of animals were slaughtered to try to stem the disease's progress. The killings continued even when the spread of the disease slowed and despite the demands of animal rights activists that the policy should be ended. But then came Phoenix. Phoenix was a cute ten-day-old calf – described by one newspaper as 'the angel-white product of an immaculate (or at least artificial) conception' – who had miraculously survived the slaughter of all the other cattle on her farm, including her mother. Despite being completely healthy, the officials from the Ministry of Agriculture, Fisheries and Food were adamant that she must die.[30] Luckily for Phoenix, a photographer arrived on the farm and took her picture – a pure white calf peering curiously out of the black background of a shed like a cuddly toy. The picture eventually made it to the national press and soon a campaign to save Phoenix was in full progress. A general election was looming and the Labour government, realising that killing Phoenix would do little to keep its re-election chances alive, soon reversed its mass-slaughtering policy, causing one tabloid to warn its readers: 'Vote Labour or the calf gets it!'

There have been similar cases of public emotions being aroused by reports of the plight of individual animals. In 2002, £61,000 was spent rescuing a dog called Forgea that had been abandoned on an oil tanker that was adrift in the North Pacific Ocean after an explosion. In Holland there was outcry in 2005 when a house sparrow was shot after it had flown into a building where preparations were being made for an attempt on the world domino record. The bird had landed on a few dominoes and caused 23,000 others to topple. A website, set up as a tribute to the dead sparrow, received condolences from around the world. Today the stuffed bird resides in the Natural History Museum in Rotterdam perched triumphantly on a red box of dominoes.

What happens when emotive stories and images appear alongside statistics? It appears that the numbers serve only to *dilute* the emotional impact, reducing people's compassion. Deborah Small, a professor of marketing and psychology at the University of Pennsylvania, ran an experiment where participants were invited to donate money to the Save the Children Fund.[31] One group saw a picture of Rokia, a 7-year-old girl from Mali, Africa, who was described as desperately poor and facing a threat of severe hunger or even starvation. A second group was only given statistical information, such as 'Food shortages in Malawi are affecting more than 3 million children'. A third group received both the picture and description of Rokia and the statistical information. As expected, the donations from those who only saw the statistical information were less than half those received when just the description and image of Rokia were provided. But, surprisingly, adding the statistical information to Rokia's image and description also decreased donations by almost 40 per cent. It appeared that the statistics lessened people's reliance on their emotions to guide their decision. Instead they fostered a more analytical mode of thinking and this caused the participants to donate less. For example, they may have reasoned that, faced with a tragedy on such a large scale, their donation would be a mere drop in the ocean and could make little difference.

We're the exception

Ryan (not his real name) had always dreamed of being a successful entrepreneur. He had already had a few small-scale successes buying items in Sunday car boot sales and selling them at a substantial profit on the online auction site eBay. So when he got a message from the producers of BBC's long-running show *The Apprentice* that he'd made the next round of candidate selection, he had a decision to make. The selection was due to take place in London. He was working in a remote town in Western Australia, serving breakfast and cleaning rooms in a small hotel. Was it worth the expensive 9000 mile trip back to the UK? Ryan didn't think twice. Sharp and creative, he was buzzing with ideas. And the opportunity to win the show and go into business with the show's acerbic star, Lord Sugar – one of Britain's most famous and successful businessmen – was unmissable. Ryan borrowed money and booked his flight to Heathrow.

But there were a few things that Ryan didn't know about the selection procedure. One is that it pays to be obnoxious. That's as important as performing well in the mock business-related tasks that the candidates are asked to carry out. The producers wanted big, loud, arrogant personalities who would shout, rant and argue with their rivals to avoid the blame if their team came second-best on a particular round.[32] That makes good television. Ryan made the mistake of focusing on impressing the judges with his business acumen.

However, there was something else Ryan could have considered, even before he booked his flight from Perth. He was one of hundreds of candidates who would be invited to attend that stage of the selection. Eventually, only sixteen would appear in the series and one would be the winner. The odds were massively against him. Sure enough – jet-lagged, in debt and now jobless – he left the selection venue disappointed.

What Ryan hadn't considered was what psychologists refer to as the base rate – the probability that, all else remaining equal, he would be selected for the show and even win it. If there were, say,

500 candidates who made it to his stage of the selection process, then he only had a 3.2 per cent chance of getting on the show, and a tiny 0.2 per cent chance of winning. Of course, we shouldn't condemn optimism. There are many examples where people have achieved great things despite the odds. Many leading businesses would never have got off the ground if the decision on whether to launch them had been based solely on their statistical chances of success. Nevertheless, it's still wise to be aware of the risks we are facing so that we can make a considered decision and possibly make contingency plans in case things go wrong. So why do people often pay so little attention to base rates?

One reason is that, as we saw above, we are often attracted more to colourful stories and anecdotes than to cold statistics. But there is another. We tend to focus on the specific details that relate to ourselves, rather than stepping back and taking a wider view of how well people similar to ourselves tend to fare. Ryan focused on his own attributes as a potentially successful entrepreneur. This told him that he must have a good chance of getting on the show. He took what psychologists call an 'inside view'. An 'outside view' would have told him there were hundreds like him who would not be selected.

With an inside view, because we concentrate on details relating to ourselves, we tend to see ourselves as an exception that will buck the trend. In the USA the divorce rate for third marriages is between 70 and 73 per cent,[33] but that did not deter over 9 million Americans who have been married at least three times according to 2013 census data. In London, only half of businesses started in 2013 survived for at least three years, but new businesses continue to appear.[34] A study by Gavin Cassar, professor of accounting and control at INSEAD, found that, ironically, prospective entrepreneurs' optimism can arise when they sit down and produce detailed financial projections. As a result, they focus on the specifics and details of their particular venture and how barriers to success might be averted, that is, the financial exercise fosters an inside view.[35]

The costs of taking an inside view are particularly evident in large-scale projects such as the Scottish Parliament at Holyrood, the Sydney Opera House, the Channel Tunnel, the Humber Bridge, Concorde, the British Library, the 1976 Montreal Summer Olympics, and London's Jubilee Line extension. In all these cases, construction costs exceeded the initial estimates by eye-watering amounts. For example, the cost of building the Scottish Parliament was originally estimated to be about £40 million, but by the time the project was completed the bill had soared to £414 million. Similarly, the Sydney Opera house was completed at fourteen times over budget. The problem is that planners see each project as unique and fail to look at how similar projects fared compared to their budgets. As the economic geographer Bent Flyvbjerg of Oxford's Saïd Business School, argues: 'The thought of going out and gathering simple statistics about related projects seldom enters a planner's mind. Planners may consider building a subway and building an opera house to be completely different undertakings with little to gain from each other. In fact the two may be – and often are – quite similar in statistical terms, for example as regards the size of cost overruns.'[36]

We're bulletproof – we don't need numbers

The tendency to overlook base rates can be exacerbated when people have the power to influence and control other people and resources, according to psychologists Mario Weick of Durham University and Ana Guinote of University College London. Powerful people experience a life where they are less prone to external circumstances and constraints on what they can do than the relatively powerless. As a result, they tend to focus on factors that they see as central to achieving their aims and they are likely to be confident that they can overcome obstacles. Information on external factors, such as potential threats, therefore receives much less attention; statistics telling them how other events played out are seen as peripheral and irrelevant.[37]

When it comes to generating a sense of power over events – in reality, an illusion of power – there is little to rival groupthink (see Chapter 1). Groupthink, first described by Yale psychologist Irving Janis, can occur when people meet to make important decisions, often under conditions of stress and pressure.[38] If the group is cohesive – consisting of friendly colleagues who have worked together for a long time, for example – and if there is an imperious leader, there is likely to be little debate comparing different courses of action. No one wants to rock the boat. To maintain harmony, group members feel a pressure to conform with what appears to be the prevailing view. As each member provides further rationalisations reinforcing the current position, a spirit of excessive optimism can prevail, together with a willingness to take extreme risks. In these circumstances, the group has no motivation to seek out further information that might discourage its decisions, and any available evidence that conflicts with the group's choice is likely to be ignored or discounted anyway.

In his early studies, Janis found evidence of groupthink in the Kennedy administration's calamitous decision to sponsor an invasion of Cuba at the Bay of Pigs in 1961 by an army of Cuban exiles. Objections to the plan and evidence of its riskiness were dismissed by the close-knit group around Kennedy, while senior advisors within the group kept their concerns to themselves. As a result, there was a failure to gather reliable intelligence on the strength of Cuban forces. Convinced of the weakness of the Cuban defenders, and certain that the island's population would rise up against Fidel Castro's communist regime as soon as the attack began, the administration provided insufficient air cover and support to the invading force. Within three days, the attempt to dislodge Castro had failed and 1200 invaders had been captured. 'How could I have been so stupid?' Kennedy asked himself after the debacle. It was presumably little comfort that he hadn't been alone.

Since then, many other examples of the ruinous effects of groupthink have been identified.[39] These include: the doomed decision by NASA to launch the *Challenger* space shuttle in January

1986, Margaret Thatcher's insistence in 1990 that local government funding in the UK would be replaced by a poll tax, the collapse of Swissair in 2002, the damaging decisions made by Marks and Spencer's directors in the early 2000s, and Tony Blair's decision to take Britain into a war with Iraq in 2003.

So, discounting accurate numeric information, or failing to gather it, can be perilous. It can lead to societies ruled by uninformed prejudices. It can limit our response to tragedies. It can lead to rash decisions and disasters. And it can leave us naively exposed to extreme risks.

But while erecting barriers to numbers may cause us to blithely take unjustified risks in some contexts, it can also cause us to be overly risk-averse in others. In the next chapter we'll look at the angst that pervades many Western societies, despite the fact that in many ways the world has never been safer.

10

Safety in Numbers

Anxious times

Some world events are so big that we all remember exactly where we were and what we were doing when we heard the news. Over half a century later, I can still recall settling down for ITV's Friday night quiz show *Take Your Pick,* with 'your quiz inquisitor', the towering Michael Miles, and its star prizes of gleaming early-sixties cars. Then came the shocking interruption: President Kennedy has been shot in Dallas, Texas. I rushed to tell my mother, wreathed in steam in the kitchen as she transferred the week's washing from one tub to another. *Take Your Pick* resumed, but half an hour later the hospital soap opera *Emergency Ward Ten* was abandoned – the president had died. Even at that young age, I felt a frisson of fear. Those huge stable structures that held the world together appeared to have been weakened. Somewhere out there, I sensed, an indefinable threat was lurking.

Nearly forty years later I had a similar feeling. One afternoon, returning from a mundane trip to purchase replacement tyres for the car, I decided put the TV on while having a coffee. It was an odd time to be showing a disaster movie, I thought. Smoke was pouring from a distant skyscraper, and terrified and astonished voices could be heard from those watching. Then I realised that this was a special news broadcast. This was reality. Airliners had been deliberately crashed into the twin towers of the World Trade Center in New York. This was 9/11.

By 2001, television technology had developed enormously since Telstar, the pioneering satellite, relayed its low-resolution monochrome pictures of Kennedy's funeral across the Atlantic. But, even then, the delayed images of the assassination itself caused the author Anthony Burgess to write:[1] 'We have seen everything now; that impartial eye has looked on murder; from now on there will always be the stain of a corpse on the living-room hearthrug.' And now, in a corner of my living room, a disaster was following its terrible and unpredictable course, live and in colour. Outside, the trees swayed gently in a late summer breeze. Inside, an inferno, albeit more than three thousand miles away, was raging. It was impossible to turn the TV off, impossible to forget those vivid awful scenes.

With events like these delivered straight into our homes, it's perhaps not surprising that many people see the world as a fearful place. Nearly two decades after 9/11, America is still largely a society riven with anxiety.[2] Fear of immigrants, fear of economic collapse and unemployment, fear of Russia, fear of terrorism, and countless other worries, helped to propel Donald Trump into the White House. But these days, it's not just Americans who live in an age of anxiety. In late 2018 German angst was highlighted in an exhibition at the German history museum in Bonn with the title, 'Fear: A German State of Mind?' In Britain, a survey of 2000 adults conducted in 2009 by the Mental Health Foundation found that 77 per cent of people thought the world was more frightening than in 1999. The IT firm Unisys even produces a 'global fear index' (officially called the Unisys Security Index), based on surveys of people in thirteen countries, including the US, UK, Australia, the Philippines and Argentina. This index, which is designed to reflect people's concerns about their national and personal security, rose by 20 per cent between 2014 and 2017 to its highest level since it was introduced in 2007.

And yet, a look at the numbers suggests that the world has never been a safer place. 'This is probably the most peaceful time in human history,' claims Harvard psychologist Steven Pinker.[3]

Since the end of the Cold War in the late 1980s, there has been a 40 per cent decline in armed conflicts.[4] In America, violent crime has decreased by 70 per cent since the 1990s.[5] The Crime Survey for England and Wales indicated that the number of offences had decreased by over 68 per cent in the ten years up to March 2016. In many countries there have been remarkable improvements in road safety over the last forty years, despite the surge in car ownership and the relentless rise in the tonnage of freight carried on the roads. Transport specialists attribute this to better enforcement of laws, better education of drivers, and better road engineering.[6]

Not only are we safer, on average, we are also far more prosperous. Global gross domestic product per capita rose from nearly $7800 in 1980 to over $14,700 in 2015.[7] In the USA, over the same period, it increased from nearly $30,000 to over $52,000. Alongside growing economies, life expectancy at birth has soared in most countries. In a century, India's life expectancy has tripled while South Korea's has quadrupled.[8] In the UK, in the thirty-five years from 1980, it increased from 73.7 to 81.6.[9] And technology offers exciting prospects for our future health care. It's likely that robots will increasingly be available to carry out operations with superhuman precision, while gene editing will lead to the elimination of many life-limiting genetic diseases. Some scientists even envisage the imminent development of anti-ageing pills that will comfortably sustain us in active lives as centenarians.

Even global deaths from natural disasters have plummeted in the last hundred years. In the 1920s, there were about 28 deaths for each 100,000 people in tragedies caused by natural phenomena like earthquakes and storms, with droughts, in particular, leading to major losses of life. Between 2010 and 2015 these forces were responsible for only about one death per 100,000 people, earthquakes now being the major cause of death.[10] Again, technology can be thanked for this astonishing improvement. We can now predict hazards much more accurately, giving people time to flee from dangers. Once a disaster has occurred, modern transport and distribution networks enable essential resources such as food or hi-tech rescue equipment to be delivered to affected areas much faster than was possible a century ago.

So why aren't we reassured by statistics like these? Why do we ignore or misinterpret the good-news stories painted by the numbers? And why do we readily accept significant risks while worrying about threats that statisticians tell us aren't there? We'll see later that our distrust of experts can cause us to dismiss reassuring statistics, particularly where new technology is involved. But as we'll also see, there are others who stand to gain if they can convince us that the world is a scary place. And the result can be an unholy alliance between vested interests, the modern media, and the way our brains function. The anxieties that emerge from this nexus not only reduce our sense of well-being and the quality of our lives, they can also subject us to greater dangers and lead to huge misallocations of public money.

Marketing anxiety

For more than six months, Norman (not his real name), a relation of mine, had looked forward to the holiday in Turkey, organised by the British Legion, that he was due to take with his wife. An energetic and engaging 92-year-old, this would be his first trip to the East since he'd served in the air force in India during the Second World War. India in those far-off years had seemed racked with poverty and disease, its oppressive humidity relieved only when the monsoon arrived. He remembered standing fully clothed in the deluge, desperate to cool down, unwilling to move as his saturated shirt and trousers tightened around his body. But Turkey in April would be different. A modern country with a booming economy, the ancient bazaars in Istanbul still promised an enchanting oriental tapestry of colour, smells and sounds that he'd never found in wartime India.

'This might be the last time we'll go abroad for a holiday,' Norman told me, his hand caressing the top of his head in a vain attempt to smooth hair that wasn't there, 'so we're really excited about it.' For Norman and his wife it was a chance to see what the world beyond the Balkan mountains was really like, a chance to experience the wider world before they considered themselves too old to travel any more.

But the next time I spoke to Norman he was on the verge of cancelling the trip. It was December 2016 and two explosions – a car bombing and a suicide bombing – in Istanbul's Beşiktaş district had killed 48 people; another 166 were injured. A Kurdish separatist group claimed responsibility. 'It would be daft to put ourselves in danger,' Norman sighed, glancing at the BBC's rolling news coverage. I sensed that the TV had been tuned to that channel all day. 'We've already paid for the trip and I don't expect to get a refund.'

It was time for me to climb on to my high horse. 'Statistically, you'll be in much more danger driving to Heathrow airport,' I began. 'Terrorists thrive on making us all feel unsafe, when in reality the risk they pose to any one of us is extraordinarily low. Your own kitchen, with its hot surfaces, boiling water, gas, electricity and sharp knives is probably more dangerous than the streets of Istanbul. You go ahead and enjoy your holiday. You've both earned it.'

Norman had always had an interest in science, so I presumed he'd be receptive to rational arguments. 'Of course, what you're saying makes sense,' he said, his hand rubbing his eyebrows to signal his dilemma, 'but ...'

I sought to press my advantage home. 'In 2015, 0.0004 per cent of people died in terrorist attacks worldwide.' I'd recently read that figure somewhere and it had stayed with me. But rather than impressing Norman, it had the opposite effect.

'Figures like that just don't register with me: 0.4 per cent, 0.04 per cent, 0.000004 per cent, it all sounds the same to me. What does 0.0004 per cent look like?'

I attempted to resort to humour, though it crossed my mind that what I was saying didn't quite fit the argument I was trying to make. 'Look, statistics show that more people die in bed than any other place, so I hope you've been sleeping on the floorboards for the last few years.'

Norman cancelled the holiday. He told everyone else that he couldn't travel because he had a bad back. I did check with his wife that this hadn't been caused by him sleeping on floorboards and was reassured, with a quizzical look, that it wasn't. Surprisingly, Norman's back was back to normal the day after he received a partial refund for the holiday. There goes another victim of irrational thinking, I thought. If only people would look at the statistics and base their decisions on facts.

But then, a couple of months later, I was walking in the countryside beside a pleasant stream in early spring. I couldn't resist a quick glance at my phone. It alerted me to the news that someone had deliberately driven into pedestrians on Westminster Bridge before fatally attacking a policeman outside the Palace of Westminster. The authorities declared this to be a terrorist attack. I was due to go to London two days later and intended to walk across that bridge. Suddenly, that cool rationality that I had tried to impose on Norman deserted me. Would I be safe? 'I wouldn't go to London,' a neighbour advised me.

The American neuroscientist David Eagleman of Stanford University has characterised our brains as a parliament, with different parties presenting competing arguments as they battle to determine our behaviour.[11] One party reminded me of the minuscule risk posed by a terrorist attack; that I encountered small, but greater, risks every day without giving them a moment's thought. But there was still an unsettling voice from the backbenches that warned me that London was unsafe. Neuroscience indicates that this voice actually comes from the stria terminalis, a pathway in the brain that connects the tiny almond-shaped amygdala to the hormone-secreting hypothalamus. Fear – how I'd react to an immediate threat like someone pointing a gun at me – appears to be different from anxiety, which is a worry about a threat that just might be lurking in the future.[12] In other words, anxiety arises from uncertainty and it is the stria terminalis that is automatically activated in response. So, by sowing uncertainty, those with ulterior motives can easily create anxiety, even if the statistics tell me to relax. Evolution has designed us to prioritise anxiety. The sound of a twig cracking in a dark prehistoric forest might have been benign, but while running away unnecessarily would have wasted a few valuable calories, ignoring the threat could have cost us our lives.

There are plenty of people with motives to trade on uncertainty and anxiety, and they are not all terrorists. Public anxiety can be manna to a particular variety of politician, particularly those at the far conservative end of the spectrum. Demagogues offer simple authoritarian solutions to people's worries: the pain relief of certainty, usually coated in an easily swallowed capsule of villains and scapegoats. Research by George Bonanno of Columbia University and John Jost of New York University suggests that politicians like this are more attractive to those who have heightened perceptions of the world as a place with unbearable levels of crime and terrorism.[13]

Frightening stories also bring readers and viewers to the media. The effect is to amplify the rare, but scary, events that occur around the world. The modern news machine targets us twenty-four hours a day and directly to our mobile devices – saturation coverage that makes the infrequent seem like the norm. Pharmaceutical, health and insurance companies also have a vested interest in playing on our worries, sometimes abetted by the media. One website has collated 170 things that can give people cancer – ranging from air travel and aspirins to Wi-Fi and wine – according to *Daily Mail* reports over the years.[14]

Dramatic, recent and vivid events, and stories highlighted in the media like terrorist attacks, plane crashes and nuclear accidents, are seared into our memories, but we tend to forget the humdrum. As the eminent psychologists Amos Tversky and Daniel Kahneman (now a Nobel laureate) showed, this gives us a distorted perception of the world because the ease with which potential threats are brought to mind becomes a rule of thumb for assessing the chances that they will materialise. A rare train crash that dominated the headlines last week creates the suspicion that rail travel might be unsafe. Television pictures of an octogenarian, injured after being attacked in their home by an intruder, causes fear among the elderly even though, fortunately, such events are extremely uncommon.

The numbers show that it's the mundane threats – food poisoning, diabetes, a fall in the street, smoking, influenza and obesity – that are more likely to dispatch us than train crashes or violent crime. But these insidious killers rarely make the headlines. Not a single passenger died on Britain's railways as a result of a train accident in the eight years up to September 2015.[15] In America, you are 36 times more likely to die of heart disease than to be murdered,[16] yet I remember being worried about a trip to Washington DC after an American I met at London airport told me about the high levels of gun crime in the city. Sharks evoke terror in many people, and when they attack humans, the news cascades around the world. The film *Jaws* owed much of its success to our primitive fear of sharks, yet the US, for example, averages one fatal attack every two years.[17]

When our perceptions of risk deviate significantly from what the numbers tell us, we can even put ourselves in more danger by attempting to avoid what is, in reality, a tiny threat. In the immediate aftermath of 9/11, many people were, understandably, worried about flying, so they switched to cars. But flying is much safer than driving, even after considering the risk of a plane being downed by terrorists. David Myers, a social psychologist at Hope College in Michigan, estimated that, even if terrorists hijacked and crashed one plane each week for a year in the USA, with each plane carrying 60 passengers, it would still have been safer to fly rather than drive.[18] After 9/11 more than 1500 people are thought to have died in road accidents because they took to their cars to avoid the minuscule risk of being a direct victim of aviation terrorism.[19] That's six times the number of people who were aboard the four fatal flights. A similar phenomenon was observed in London following the 7/7 attacks in 2005, when some people switched from travelling on the underground – where a bomb had been detonated on a train – to cycling. Sadly, cycling through London's traffic is a relatively hazardous pursuit. Peter Ayton, a psychologist at City University, estimated that there were 214 more casualties in the six months after the bombing than would normally have been expected, though few of these were fatalities.[20]

Making decisions that increase the dangers we face is not confined to our transport choices. After newspaper headlines about a child's abduction, anxious parents often restrict their children's freedom, with potentially negative effects on their physical health, despite the extreme rarity of such crimes. Some parents have even considered having GPS tracker chips inserted under their child's skin so that their location can be constantly monitored. The operation needed to implant such a device is not risk-free.

Governments are not exempt from decisions that ignore relative risks. In Britain in the early 2000s, annual deaths on the roads were around 330 times those on the railway system. After two major rail crashes in the London area, in Southall in 1997 and near Paddington in 1999, a 'Joint Safety Enquiry' recommended that an Automatic Train Protection (ATP) system should be installed across the rail network. The system automatically stops a train from passing a red signal; it was estimated that this would save an average of two lives a year at a cost of about £2 billion. This was approximately 200 times the amount spent on preventing a road death.[21] As Kevin Delaney, spokesman for the RAC Foundation, said at the time: 'The trouble is that we have a rail crash and 20 or 30 people are killed, whereas there may be ten road deaths a day but they don't happen in the same place at the same time.'[22] In other words, the road deaths don't usually make the front page of national newspapers, but a train crash will almost inevitably be accompanied by headlines asking: Are our railways safe? The answer: They are very safe indeed.

'I see you got back from London in one piece,' shouted my neighbour, interrupting his lawn mowing, a week or so after I'd returned. For a few moments I wondered what he was taking about. By then the news had returned to its habitual coverage of Donald Trump, who'd decided to drop 'the mother of all bombs' in

Afghanistan, while North Korea had tested another nuclear missile and promised to continue doing so on a weekly basis. I realised, with surprise, that I'd crossed Westminster Bridge thinking only about the meeting I was attending and without a second's thought about the events of a couple of days before. London had seemed normal – intense, noisy, crowded, rushed and, for a provincial like me, exciting.

'*I* wouldn't have gone,' he said, hinting that I'd been reckless.

'It's all in the mind,' I replied. But I don't think that he understood me.

Ten per cent of very little is very little

Rather than persuading us to ignore numbers about risk, by highlighting stories about very unlikely threats, those with a vested interest in making us worried sometimes use incomplete numbers to convince us that we are probably doomed. But even if you are told that your behaviour is associated with a 200 per cent increase in risk, you may still be very safe. 'Ten per cent of nothing is nothing,' said the Canadian trades union leader Jerry Dias in 2017, when complaining about the recent pay rise given to Mexican workers whose low wages, he feared, were leading to unfair competition with his Canadian members. He could, more accurately, have said: Ten per cent of very little is very little. And statements like that should be writ large when we read reports about risk.

One day, Michael, a friend of mine, seemed concerned. 'Have you seen this?' he asked, handing me a folded copy of a newspaper before he climbed into my car.

I glanced at the headline: 'Living near a busy road may raise risk of dementia, major study into pollution finds'.[23] The increased risk was estimated to be 7 per cent. We were parked perilously outside Michael's house. Vans, cars and growling juggernauts squeezed past us in an endless procession.

I felt sorry for Michael as we set off for our day's hiking in the nearby hills. I knew he was under pressure both financially and emotionally because he had a close relative in an expensive care home suffering from dementia. And now he was being told that the very air he breathed in his own home was very likely to be damaging his brain. Every time he forgot someone's name, every time the right word wouldn't tumble effortlessly into a sentence, and every instance of absent-mindedness could now be seen as a sign that the traffic fumes and noise were having their sinister effect. As I turned the headline round and round in my thoughts, I began to see Michael as doomed. After all, this was a *major* study reported in a respected newspaper. He and his wife had lived in their house, inhaling the invisible poison, for twenty years.

'Let me read that article,' I said as I waited in the car park for Michael to struggle into his hiking boots and gaiters. I had been deliberately taking in deep breaths of fresh hill-top air in case the short exposure to the fumes in Michael's road had already put me in danger.

The Canadian study reported by the paper was impressive. The researchers had followed the medical records of every person aged between 20 and 85 living in Ontario over the years 2001 to 2012. They had used postcodes to determine how far they lived from a busy road. Many experts commented that the study was important and that its findings were highly plausible, though it was unclear whether fumes or traffic noise or both were the cause of the increased risk. Some pointed out that other parts of the world experience far worse cases of traffic pollution than Ontario, so the increased risks elsewhere might be even higher.

But what were the increased risks? In the fourteenth paragraph of the article were those boring, but scientifically precise, figures we often skate over. If you lived within 50 metres of a main road, like Michael, you had a 7 per cent increased risk of developing dementia. For those living between 50 and 100 metres from a main road, the risk increased by 4 per cent, while it was only 2 per cent for those living between 101 and 200 metres. Beyond this range there was no increased risk at all.

What was missing from the article was an indication of the risk of dementia if you don't live near a busy road. The 7 per cent was an increase from what? Some sources estimate that the risk of getting dementia, wherever you live in the Western world, is about 11 per cent. Seven per cent of 11 per cent is a tiny 0.77 per cent. So, according to the statistics: take 100 people who live anywhere and typically 11 will eventually have dementia. Take 100 people who live very close to a main road and just under 12 will develop the condition, on average. This didn't sound quite as scary to me as the headline.

Of course, we should not discount the prospect of saving those extra lives. Measures such as ensuring new homes built close to busy roads have their lounges at the back of the house are well worth considering, while it makes sense to intensify efforts to reduce traffic pollution. This is especially true as air pollution carries other hazards such as aggravated respiratory conditions and heart disease. But exaggerating dangers by reporting relative risks (the 7 per cent increase in risk) rather than absolute risks (the actual risks you face: 11.77 per cent versus 11 per cent) creates unnecessary alarm. Almost weekly, we are assailed with stories couched in the same terms. We are told a daily small glass of wine can increase the risk of breast cancer in premenopausal women by 5 per cent,[24] that loneliness may increase your chance of a heart attack by 29 per cent,[25] and, more happily, that eating prunes reduces the risk of type 2 diabetes by 11 per cent.[26] But if the reports don't tell us what the absolute risk is then the numbers are almost meaningless.

X causes the dreaded Y

Should we be worried when the fashionistas decide that the next season's women's dresses should have hemlines below the ankle? Long dresses tend to mean that economic recessions are on the way, while short dresses herald an economic bonanza. As the stock market boomed in the early 1920s – the age of the flappers – women's skirts shortened. Then came the Great Crash of 1929 and, sure enough, hemlines lengthened. Fast forward to the swinging sixties and, as a rapidly growing economy allowed us to fill our houses with the latest consumer goods, girls paraded the trendy streets of London in miniskirts. This oscillating partnership between hemlines and our economic fortunes has persisted through the oil crisis and stagflation of the 1970s, the yuppy boom of the 1980s, and beyond. And this is not an illusion based on selective memory. One serious academic study found that the correlation between hemlines and US stock market prices from 1929 to 1970 was highly statistically significant, meaning that it was unlikely to have occurred by chance.[27] Why, oh why, doesn't someone from the government tell those fashion designers to keep those hemlines high, I wonder. If they did that, there'd be no need for quantitative easing, unbalanced budgets, or bargain-basement interest rates.

Of course, I am making the mistake of inferring a causal link where there is none. Just because two things happen to move up and down together in tandem does not mean that one is causing the other. In this case, there *may* be a causal link in the opposite direction: the state of the economy influences skirt lengths, rather than the reverse, though why this would be the case isn't immediately obvious.

The concern is that unquestioningly assuming causality can mislead us into blaming innocuous factors for the dangers we face. Take these scary headlines: 'Spanking naughty children increases their risk of depression and substance abuse';[28] 'Snoring linked to Alzheimer's'; 'Women and men who regularly trim or remove all their pubic hair run a greater risk of sexually transmitted infections'; 'Netflix and kill warning as watching too much telly can increase risk of dying';[29]; 'Tooth loss link to increased risk of dementia'. Reading these might make you feel you are living in a deadly pinball machine, triggering a dreadful risk every time you move. But in none of these reports was a causal link established between the apparent menace and its grim outcome. The researchers involved in the studies usually presented a theory as to why one thing was causing another, but as we saw earlier, we are brilliant at inventing reasons why X might be causing Y even when, in reality, they are unrelated. This is not to say there were no causal links, but only that they were far from proven.

There are several reasons why X might not be causing Y even though they occur or change together. Pure coincidence is one explanation. The population of the USA has increased since the Second World War and so has the number of miles of restored canals in the UK, but that's not because thousands of Americans crossed the Atlantic to escape their overcrowded cities to dig out the old waterways. When researchers base their conclusion on very small samples of people these coincidences are more likely to occur. For example, a study suggesting that tattoos are associated with a greater risk of suffering from heatstroke was based on just ten male participants.[30] However, some coincidental correlations survive even when data on millions of people are gathered. If you collect data on enough variables – people's height, their weekly shopping expenditure, the time they spend travelling to work, whether or not they are vegans, the number of siblings their father had, whether they've had a cold in the last month, and so on – some of these are bound to be correlated even though they are unrelated. This is a problem in an era of big data. We are in danger of being presented with so many potential causal relationships that are spurious that we miss the ones that matter. Some of the relationships that have been uncovered are merely amusing: Facebook analysts have found that, if your name is Yvette, you are 37 per cent more likely than the average person to have a sister named Yvonne. Others make good news stories, but are merely frivolous: per capita consumption of margarine in Maine is closely correlated with the state's divorce rate (but, more pleasingly, the number of honeybee colonies in the USA is highly correlated with the marriage rate in Vermont).[31] Some correlations might just hint at some hidden trait of human behaviour, but we can't be sure. For example, one US study found that black vehicles are less likely to be purchased when the weather is warm and sunny.[32]

Sometimes spurious correlations occur because the two factors being examined are both influenced by a hidden third factor. In the case where smacking naughty children was found to be correlated with the risk of the child's substance abuse in later life, this factor might be the lifestyle of the parents. Parents who are themselves substance abusers may be more likely to smack their children, while having parents who are substance abusers may itself increase the chance that their children will develop similar problems. The smacking itself might have no direct influence on the chances of substance abuse. Also, the study was based on adults' ability to recall whether they were smacked. It's possible that people with mental health problems, such as depression, have a greater tendency to recall negative childhood experiences.[33]

Similarly, sexual activity might be the hidden factor that leads to the correlation between pubic hair grooming and the risk of contracting a sexually transmitted disease. The researchers in this study speculated that the grooming process could lead to small skin lesions that increased exposure to viruses and bacteria. However, it may simply be that more sexually active people are more likely to indulge in grooming and also more likely to be exposed to infections.

In other cases, as we've seen, there may be causality, but it might work in the opposite direction to that which is reported. The *Daily Express* claimed that brushing your teeth will help to protect you from dementia, as those with tooth loss were found to be more likely to suffer from the condition. But it may be that it's the dementia that causes tooth loss rather than the other way around, because people with disorientation and memory loss may forget to brush their teeth regularly and may have a poorer diet.

Despite these limitations, people can be too ready to assume causality based on dubious evidence. The website naturalhealth365.com claims: 'Root canals and breast cancer: The connection is clear'. It refers to a study of 300 women with breast cancer by Dr Robert Jones. He found that 93 per cent of them had had root canal surgery. Of course, this in itself provides no evidence of causality at all. It's probable that around 93 per cent of the women were also meat eaters, that around 93 per cent of them had access to a mobile phone, or that around 93 per cent were less than 70 inches tall.

The problem is that, having established a belief that one thing causes another, we tend only to recall cases that confirm this belief. Disconfirming evidence has less chance of registering with us. 'Yes, that figures,' we might say to ourselves. 'Mrs A. had root canal surgery and she suffered from breast cancer.' The many people who have breast cancer but have not undergone root canal surgery, or those that had root canal surgery but did not suffer from the disease, go unrecognised. The result: we cancel a dental appointment that would have saved us lots of discomfort for no good reason at all.

We don't trust those experts

We can't blame all of our worries about tiny risks on scaremongering reporters and politicians or profiteering pharmaceutical companies. Sometimes we just have an instinctive fear of things that are new, unfamiliar, or outside our control. And no amount of reassuring statistics will be sufficient to make us feel safe.

Short-haired and in her thirties, Hayley stands firm and unsmiling for the photographer from the local newspaper, a baby twin in each arm and a toddler by her side. He seems distracted and amused by some event in the middle distance. Behind her, a grassy open space leads to the shade of sycamore trees in full leaf. Just beyond, there's a glimpse of modern houses and a touring caravan. It looks as if the photographer called on Hayley unexpectedly. She is still wearing her slippers. She looks like an unlikely foe for an international corporation with a turnover of 47.6 million euros. But somehow her power lies in her vulnerability. Who would dare to risk hurting a family like this?

The perceived threat to Hayley's family comes from Vodafone. It wants to erect a fifteen-metre phone mast on the ground in front of her home. It will come with two microwave dishes and Hayley is concerned that it could increase the risk of her children developing childhood leukaemia or brain cancer. She also points out that there is a school less than 200 metres from where the mast would be sited.

The work of the US psychologist Paul Slovic and his colleagues suggests that mobile phone masts have many of the features that cause the public to regard risks as unacceptable even when statistics suggest that they are tiny or non-existent. As individuals, we have little control over where masts are sited; they are imposed on us involuntarily and they represent a relatively new technology that works in ways that are a mystery to most of us.

Things we think we can control present more acceptable risks than potential threats judged to be outside our influence. Some people are happy to ski but studiously avoid food preservatives even though leisure time spent on the piste is estimated to be a thousand times more likely to cause injury or damage to one's health.[34] Similarly, we may discount the risks of driving a car ourselves, but we would probably be scared if we were forced to be a passenger in a self-driving car, hurtling along the motorway at 70 miles per hour, a few metres behind the vehicle in front. Our fear would probably persist even if the statistics show that 90 per cent or more of car accidents can, at least in part, be blamed on human error and that accidents caused by self-driven cars are very unlikely.

We also find risks to be less acceptable if they are imposed on us involuntarily, distributed inequitably, or offer us few immediate benefits. People may protest about pollution, or the addition of fluoride to the water supply, while happily accepting the risks of smoking or high levels of alcohol consumption, two of the largest causes of preventable deaths in both the UK and USA. There is even some evidence that we voluntarily increase the risks we are exposed to in an effort to maintain a constant acceptable level of perceived risk in our lives. When things are made safer, we compensate by behaving less cautiously. One study observed the behaviour of motorists as they approached an uncontrolled railway crossing in Canada. Observations were taken before and after the removal of vegetation that had obscured a clear view of the track. With the better sightline, drivers tended to approach the crossing at higher speeds, increasing their stopping distance, so that the risk of an accident at the intersection was not reduced.[35]

Through evolution, we have been conditioned to be suspicious of the new and unfamiliar. As steam power and the telegraph spread across nineteenth-century America, the neurologist George Beard feared that they were causing a nervous disease which he called neurasthenia. He attributed its symptoms – fatigue, headaches, depression, high blood pressure and anxiety – to the increasing pace of life and more intense work pressures associated with these new technologies. Because Americans seemed to be particularly prone to the illness, it was also called Americanitis. There have even been suggestions that it became fashionable to be diagnosed with neurasthenia. It was a sign that you were wealthy, active and competitive, a quintessential American go-getter whose lifestyle involved engaging with these modern inventions. When that ubiquitous twentieth-century technology, television, arrived in our home I remember being warned that watching it for too long would give me square eyes. And, as the Americans and Russians raced to land a human on the moon, my grandmother feared the consequences of 'interfering with heavenly bodies'.

While most of us regard risks as being less acceptable if we feel we can't control them, or if they are unfamiliar or imposed on us, experts tend to judge acceptability based on numbers such as mortality rates or deaths per passenger mile. This means there can be a gulf between experts and the public, and in recent years that gulf has widened.

Research suggests that to trust an expert we need not only to believe that they know what they are talking about but also to accept that they have integrity and are well intentioned.[36] Of course, these qualities are not guaranteed and statistics can be easily manipulated by those with devious intentions, so it's healthy to be sceptical. For years the Lancashire asbestos manufacturer Turner and Newall – then the largest in the world – ran a campaign designed to discredit a truth that it had secretly known since 1961: the only really safe number of asbestos fibres in the works environment is nil. Besides portraying those urging action against the company as communists and engaging the weighty political support of the local thirty-stone Liberal MP, Cyril Smith, the company persuaded its in-house science expert, Dr John Knox, to write a paper to cast doubts that exposure to asbestos caused disease.[37] In Rochdale, the town where Turner and Newall was based, people would look at the trees and fields covered in asbestos powder and joke that they got frost all year round. In some streets in the town, it is thought that every other home lost a person to an asbestos-related illness.

Wild-haired boffins in white coats, performing their strange rituals in laboratories, may seem to be other-worldly and to have a limited understanding of our personal values and concerns. Their language, methods and publications are usually inaccessible to most of us, and sometimes we witness them locking horns in animated debates about the 'truth' that make no sense to us. So, if we have suspicions that experts have a secret agenda, such as protecting profits for the industry that employs them, or if we think they are promoting a biased view of the world to defend their particular academic perspective, our trust in their statistics evaporates.

Even when experts are honest and well intentioned, and the direct risk of the technologies they are defending are non-existent, we may still be harmed if we don't trust them. Take the phone mast that Hayley is so worried about. We know that high-power microwaves can heat substances, such as food in microwave ovens, but the radiation used in phone transmissions is a lot weaker and a person's distance from the mast reduces its effects even more. It seems likely that it's the *belief* that phone masts are dangerous, and the associated stress, that is the cause of any observed ill health, rather than the microwaves themselves. We know that throughout history people have become ill because they believe they have been cursed, and the same psychological processes may apply here. In a three-year study led by Elaine Fox at Essex University, people who said they were sensitive to radiation performed no better than chance guesswork when asked to detect whether a mast was switched on or off. Irrespective of whether a signal was really being transmitted, or whether they were wrongly told that a transmission was in progress, the number and severity of their symptoms was the same.[38] In fact, no scientific study has yet found any evidence that phone masts directly cause harm.

But this is unlikely to reassure Hayley. As she told the local paper: 'It won't be until ten to twenty years down the line that we could start to notice the implications of having a mast like this in the town. That's when the impact on health could start to become visible.'[39] So what should be the role of experts and their numbers in these difficult decisions? The arrogant expert who talks down to the public, attempting to drown out their concerns in a torrent of scientific jargon and portraying their objections as irrational and ill-informed is only likely to lead to entrenched positions. The genetically modified (GM) food company Monsanto has been accused of adopting this high-handed strategy, and it simply led to a backlash and an erosion of confidence in the safety of GM products. In particular, Monsanto argued that, because GM food is perfectly safe, it should not even require labelling. As we have seen, people's perceptions of risk involve more than numbers and statistics and this does not mean that these perceptions are any less valid or real. Research suggests that open and honest dialogue with people, appreciating the basis of their concerns, acknowledging any uncertainty, and allowing full disclosure of available information, is more likely to establish trust in the long term and allow people to make informed choices.[40]

After all, if Hayley's quality of life is reduced because she is worried and depressed about the perceived threat to her children from the proposed phone mast then her objections shouldn't simply be dismissed as the illusion of a woman who knows nothing about telecommunications technology. Whatever the numbers say, her distress is a reality and she has every right to expect her worries to be taken into account.

We're sinking!

The passionate man on my doorstep in a smart suit was convinced that Armageddon was just over the horizon. With each piece of irrefutable evidence that the end of the world was imminent, he banged a book he was trying to sell me against the palm of his other hand. The book appeared to take such a battering that I wouldn't have bought it anyway. Of course, the world's demise has been wrongly predicted many times, with the forecasters usually having a ready set of excuses to explain that they just needed to make a few adjustments and they would be right next time. But, after he'd left with an irritated 'You'll regret that you didn't take my warnings seriously', I was surprised to find that the tone of his message had pierced the armour of rationality that I'd presented to him. Suddenly I found myself yearning for the seemingly comfortable world of twenty to thirty years ago – a world where few of our worries had materialised and most of the threats had long since dissipated.

'Crime up by 11 per cent', 'Consumer spending falls by 0.8 per cent', 'Rail punctuality down by 1.1 per cent', 'Flu cases on the rise'. As if we didn't have enough to worry about, the latest ominous movement of the graphs tells us that the world is going to the dogs. It's not surprising that 'declinism' – the belief that the future will be worse than the past – is prevalent. In a 2015 British survey, only 5 per cent of respondents thought the world was getting better, while 71 per cent thought it was getting worse.[41]

Psychologists have suggested that two factors might lead to declinism. First, people tend to have the most vivid memories of events that occurred to them between the ages of 10 and 30 (this is called the 'reminiscence bump'). The details of events we experienced between ages 30 and 60 are less well recalled. And second, as we get older, we tend to remember positive things more readily than negative.[42] These factors combined mean that the past can seem to be a rosier place than the present, even if the numbers tell a different story.

But this perception of an ever-worsening world is not helped when we are persuaded to see negative trends that are not there.

Two swallows don't make a summer and, likewise, two numbers don't always make a trend. Many things we use to measure the state of the world are subject to random factors – unpredictable one-off events that cause graphs to bounce up and down over time. If I throw a 6 on a dice followed by a 4, it would be silly to argue that there's a downward trend in the scores. Monthly levels of crime depend on the weather (pickpockets don't like being soaked in a thunderstorm), chance opportunities that present themselves like an unlocked car, and a host of other haphazard factors. This means that if we compare this year's level of crime with the same period last year, there is bound to be a difference: last year will be better or worse than this year.

The problem is that we pay more attention when the latest movement in the graph gets worse than when it gets better, and negative news is more likely to be publicised than positive. One experiment, conducted by Stuart Soroka of the University of Michigan and Stephen McAdams of McGill University, exposed people to positive and negative television news reports and monitored a number of physiological responses that are associated with arousal and attentiveness, such as heart rate and skin conductance.[43] They found that while negative news increased both arousal and attentiveness, the physiological effect of positive news was similar to that of a blank screen. Soroka and McAdams provide evidence that magazine news-stand sales typically increase by 30 per cent when the tone of the cover is negative rather than positive. Hence, each apparent turn for the worse in a graph provides an inviting opportunity for a profitable story, even though it may simply be a random twitch. In contrast, random twitches in a positive direction are lucky if they make a short piece on an inside page.

Worst of all are needlessly depressing reports that compare successive months or quarters when there's a seasonal pattern in what is being measured. 'Train punctuality getting worse', warned a headline on the BBC News website on 12 December 2002.[44] The report was accompanied by a picture of a desperate passenger, with her head buried in her hands, in front of a station departure board. But the comparison was between the summer quarter of that year (July, August, September), when 17 per cent of trains were late, and the following quarter (October, November, December), when 20 per cent ran late. You would expect a decline in punctuality in the later months of the year when bad weather conditions can disrupt rail services, and, indeed, the report acknowledged that bad weather had been a culprit in this case. The BBC report provided no evidence at all that there was a trend of worsening punctuality. To infer the possibility of a trend we would need the evidence of several year-to-year changes, and even then there's a possibility that these are simply random movements.

Reasons to sleep at night

This chapter is not intended to suggest we should all be Pollyannas or that we should only peer at a rough old world through glasses with a rose-tinted hue. Millions of people throughout the world face lives of danger and deprivation. Global warming, pollution, international tensions and conflicts, inequality, nuclear weapons, and over-exploitation of resources present real challenges. However, we need to be clear-eyed when we assess the state of the world and its future prospects, as well as our own position within it. Ignoring numbers when they help to illuminate the truth about risk, misinterpreting numbers, or being deceived by misleading numeric indicators of danger is equivalent to wearing glasses with distorted lenses as we attempt to navigate our way through life.

Twenty-twenty vision tells us that many of the threats we fear are very unlikely to trouble us, and that the world of yesterday, which many people yearn for, was perhaps not quite as wonderful as our memories suggest. The numbers tell us that huge improvements have been made in many aspects of people's lives, yet the daily bombardment of bad news and scary tabloid headlines can blind us to these achievements. In a world where votes based on misperceptions of risk can foster misguided policies and even lead to the triumph of demagogues, our ability to access, appreciate and understand accurate numbers can be crucial.

11

When Can You Count on a Number?

True numbers

We've seen how numbers can disconnect us from the truth. We can be misinformed when we pay too much attention to numbers that, at best, only give a partial perspective on reality or when those with ulterior motives feed us with easily digested numbers they know are lies or half-truths. Numbers like these can impair the quality of some of our key decisions. We may choose to study at the wrong university, vote for the wrong political party, or pursue the wrong objective. We may change our diet on the basis of scientific results that are flawed, or have our view of the world coloured by surveys that make no attempt to represent an accurate cross section of the population. Even our self-esteem can suffer when we are labelled by simplified numbers or arbitrary categories.

A disconnection from the truth can also occur when we pay too little attention to trustworthy data. This happens when our aversion to numbers, or our inability to work out what they mean, or our prejudices cause us to dismiss reliable statistics when they have an important message for us. As a result, we may respond inadequately to victims of disasters or live less enjoyable lives because we waste our time avoiding minuscule risks.

Of course, we have assumed throughout that there is something out there that is the 'objective truth' – that Neil Armstrong did walk on the moon in 1969 and that the population of the world in 2019 exceeded 7 billion. Some philosophers dispute the notion of an objective truth. Surely, they argue, truth is subjective, socially constructed, culturally determined, and not accessible to verification. To them, one person's truth will be different from another's, depending on their background, their frame of reference, and their point of view. Not even science can provide accurate and objective knowledge; it is simply an ideology created by humans.

But those of us who are not postmodernists, who want to get as close to the objective truth as possible, face challenges. How can we tell a misleading number from a reliable one? When our intuition clashes with statistical evidence, which should we believe? And if a number is reliable, and has a crucial message, how can it be presented so that it is meaningful and has impact?

Statistical alertness

In 2015, I saw a report that claimed that 88 per cent of the US adult population had sent a sexually explicit text message ('sexted') at some time in their lives and 82 per cent had done so in the previous year.[1] These figures may be correct. Initially, I accepted them at face value. But a few days later, when talking about them to a friend, it struck me that they were incredibly high. I knew from elsewhere that nearly 14 per cent of the US *adult* population is aged 70 or over. Either these people were also sexting in large numbers or an even higher percentage of younger adults were engaged in the practice. The percentages were attention-grabbing – a gift for a reporter in search of a news story – but I had doubts.

We have more chance of dodging the hailstorms of dodgy numbers that come our way if we improve our statistical alertness. This doesn't mean you have to be a mathematician or a formula junkie. You don't need a PhD to be suspicious that a survey, sponsored by a politically slanted newspaper on attitudes to global warming or crime and punishment, is not representative of the general public, whatever the newspaper claims. A moment's reflection tells us that an average is unlikely to be typical of many members of diverse populations. Similarly, you don't need to be a trained psychologist to realise that numbers based on people's subjective judgements, like their ratings of restaurants or films, may be subject to biases and inconsistencies. Although they are potentially useful, we should regard them with caution. If we don't do this, as we saw in Chapter 9, an erroneous number can take hold and resist any attempts to dislodge it – no matter how clear and strong the contrary evidence that is subsequently presented to us.

The problem is that our System 1 (we first met this in the Introduction), which provides us with an instant, intuitive, unconscious and emotional mode of thinking, is seductively effortless. We therefore tend to default to it when assessing information and accept the first things that come to mind.[2] The cost of this ease is illusions, errors and unconscious biases. We might judge an investment in a company with an easily pronounced name, like Pera, to be more promising than one with a name like Lasiea simply because the last name involves some tongue twisting.[3] When told that a journey by road and rail took 210 minutes and the rail part of the journey was 200 minutes longer than the time spent on the road, System 1 is likely to tell us that 10 minutes was spent on the road. A few minutes of analytical thinking would tell us that the road journey lasted for only 5 minutes.[4]

This analytical mode of thinking comes from our System 2. One of its jobs is to monitor System 1 like a conscientious parent and correct it if it goes astray. But this involves relatively hard work and System 2 can be lazy, waving through substandard thinking in the hope that it won't matter. So when we are presented with a statistic, how can we jolt System 2 into action? Some techniques are not terribly attractive. Being in a bad mood apparently encourages a more analytical frame of mind, because when we feel unhappy or threatened we become more alert and cautious.[5] Frowning also activates System 2, according to one study,[6] but telling people to scowl every time they see a statistic might also do wonders for Botox sales. We are also more likely to be more analytical when we are accountable to someone else for our decisions, or when the issue we are addressing is relevant to us personally and may have consequences.[7] Telling people to look at information in the same way that a statistician would seemed to be effective in reducing bias in a German study.[8] Which raises the question of what a statistician would ask about a given set of numbers. For System 2 to be effective, it needs to be furnished with some rigorous evaluation equipment. This can involve asking about the motivation and competence of the person providing the numbers, what things the number doesn't measure, whether any comparisons are valid, and so on. A set of key questions to help you perform these interrogations can be found in the Appendix.

When can you trust your nose?

Despite the apparent cavalier nature of System 1, are there circumstances when your intuition is likely to be right after all? When the numbers tell you one thing, but your nose tells you another and your nose has it to a T? There are astonishing examples of the power of intuition. Nurses in one study were able to detect children in their care who were developing life-threatening infections, even before the results of medical tests were available, yet they could not explain how they knew the children were in danger.[9] In the poultry industry it's important to determine the sex of day-old chicks, but this is difficult if not impossible for most people even after carefully examining the birds. Yet experienced Japanese chicken-sexers can call it with 98 per cent accuracy in less than a second, based on their intuition. Like the nurses, they cannot explain how they do it.[10] Then there was the case of the Formula 1 driver who braked suddenly before a hairpin bend during a race for no apparent reason, only to find there was a pile-up of cars ahead. Had he not slowed down, he would probably have been killed. The driver could not understand his action until a team led by Gerard Hodgkinson, a psychology professor at Leeds University, showed him a video of the race. The driver realised he had unconsciously picked up the unusual reaction of the crowd.

Rather than cheering him on and waving their arms, all the spectators were transfixed and staring in the other direction, having seen the crash ahead.[11]

But consider this. Imagine you have won a competition, but the prize you've been promised is rather unusual. You will be paid a sum of money on each day for 30 days. On the first day you will received only one penny, but the amount will then continue to double each day, so you will receive two pence on the second day, four pence on the third day, and so on. It doesn't sound like it's worth collecting this prize with such tiny payments, so you tell the competition's organisers to keep their money. Then one day, feeling bored, you decide to work out what you would have won. With horror, you realise that the prize would have amounted to £10,737,418 and 23 pence. It just does not seem to make sense. How can payments that started off as small change in the first few days transform into such a life-changing sum in the short space of a month? It's a disastrous result for intuition.

So what's the difference between these examples of brilliance and ineptitude in intuition? Two factors appear to be necessary for intuition to work well.[12] First, the environment needs to be characterised by regularities. By this we mean that it exhibits patterns that can be exploited when we make our judgements. Certain clusters of symptoms must have been associated with the dangerous infections the children were exposed to, even though these couldn't be consciously observed by the nurses. The day-old chicks exhibited certain complex differences in patterns in the cloacal regions depending on whether they were male or female. And the patterns of crowd behaviour at motor-racing events will be associated with what is happening on the track.

In other areas, such as the movements of prices in stock markets, there are unlikely to be regular patterns, even though people may think they have spotted them. Indeed, stock market graphs are often said to exhibit the haphazardness of a drunkard's walk. The same lack of regularity is true of international political events, so much so that experts performed worse than random guesswork when trying to predict them in one famous study. A dart-throwing chimpanzee would have performed better, according to Philip Tetlock, the study's author.[13]

The second factor needed for intuition to work effectively is that we must have the opportunity to learn the patterns that present themselves. This requires lots of experience: we need to build a vast mental database of past cases, even though we may be unaware of its existence. Intuition has been described as nothing more or less than recognition. With plenty of practice and feedback we learn to recognise that our current situation matches earlier patterns and we automatically identify the judgement that was appropriate for these patterns. We do all this instantly and unconsciously. This means the nurses caring for the children and the Japanese chicken-sexers must have been drawing on a wealth of experience to perform so brilliantly. Had the Formula 1 driver been a novice, he might not have survived the race.

In contrast, most of us have no practice in predicting the outcome of exponential growth, such as the prize in the competition. Few things we experience increase at such a sensational rate. Our intuition therefore has no relevant database to draw on, so it makes the assumption, based on our more common experience, that growth will be slower. Receiving pence in the first few days, we surmise, must mean that the total prize will amount to very little. At least biases like this can sometimes result in a pleasant surprise. Pension pots grow faster than most people think because they earn compound interest, so the year-on-year increases get larger and larger. Intuition is likely to tell us that the growth is only linear, so we expect each increase to be the same.

In short, when numbers presented to you conflict with your intuition, your intuition may be right if you have plenty of experience in an environment that has regularities. That's precisely what happened in the early 2000s, when Bernie Madoff, the American fraudster, was caught running the biggest Ponzi scheme in history.[14] After a mere glance at Madoff's financial numbers, an experienced analyst called Harry Markopolos had the feeling that they just could not be honest. To others, with less experience and practice, Madoff's fiddling was undetectable.

Unfortunately, feeling confident that your intuition is right is no guide at all to its accuracy.[15] We might be certain that the Panama Canal is longer than the Suez Canal or that Liverpool is further west than Edinburgh, but we'd be wrong in both cases. Our confidence in our intuitive judgements tends to increase when we are given more information, but having more information can even reduce our accuracy.[16,17] Information that is consistent and has a pleasing coherence can also increase our confidence in judgements based on it, even though the numbers may be consistently biased. Psychologists call this the illusion of validity.

All of this suggests that even if our intuition feels compelling, we are more likely to get close to the truth if our two systems work in tandem. In the right circumstances our intuition can have remarkable powers in detecting anomalies, but it might also beguile us into a comfortable acceptance of misleading numbers or make us think we've found abnormalities that, in reality, are mere illusions. When an issue is important, it's therefore still best to call on System 2's analytical faculty to perform the relevant checks. That is exactly what Markopolos did. He didn't just rely on his initial intuition; this was simply the stimulus for a detailed analysis of Madoff's numbers. In this case, the analysis confirmed his suspicions.

Making useful numbers engaging and meaningful

We might perfect our cognitive apparatus to make accurate judgements about numbers that we are prepared to look at, but what about the danger of ignoring important numbers because they appear to be boring or impenetrable? When I walked into a lecture room at the start of a course, and cheerily announced that I was the statistics teacher, I could often detect an undercurrent of apprehension in the tiers of silent, fixed, unsmiling faces. I could tell what the students were thinking.

'This guy is going to bamboozle us with unfathomable formula and strange terminologies and numb us into boredom with rows of figures that have no relevance to us at all. And then he's going to expect us to pass an exam.'

The American statistician Edward Tufte once said: 'If the statistics are boring, then you've got the wrong numbers.' But as we've seen, even when we have the right numbers – honest reliable numbers that can help us to make better decisions – we often fail to engage with them or understand what they mean. So how can dry numbers be turned into messages that are impactful, meaningful and appealing?

The art of charts

We saw in Chapter 9 how art can be used to reify statistics on numbers of casualities in natural disasters and wars. But advanced artistic skills are not always needed to convey statistics effectively. Even simple diagrams, like pictograms, can work. If a patient is told by a doctor that if they don't take an anticoagulant, they have a 0.6 per cent probability of having a stroke in the next year, it's difficult to envisage what this means. In cases like this, pictograms, like the one shown below, have been found to be helpful in conveying what the probability means. This shows 1000 faces. The 994 smiley faces show the people who won't have a stroke, while the 6 paler sad faces represent the unfortunate people who will. Organisations such as NICE (the National Institute for Health and Care Excellence) in the UK are using diagrams like this to help patients to make informed decisions about their treatment.[18] Pictograms like this work, in part, because they depict numbers of people. As we saw in Chapter 7, psychologists have shown that we handle information couched in terms of numbers (or frequencies of occurrences) better than probabilities or percentages.

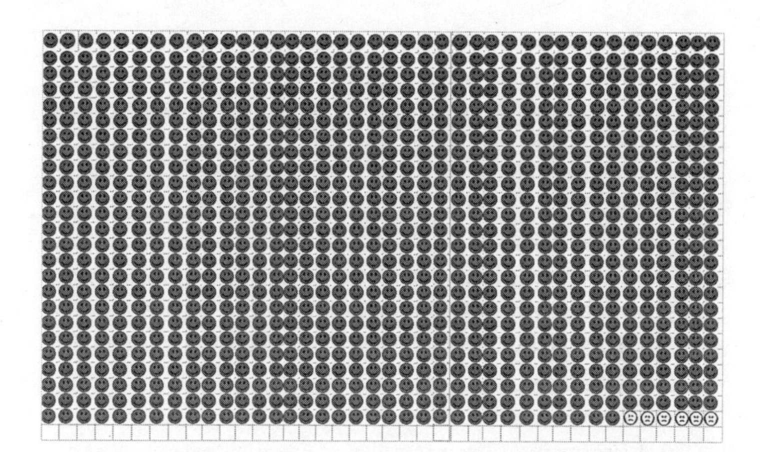

People like Edward Tufte and the late Hans Rosling, professor of international health at the Karolinska Institute in Sweden, are famous for pioneering effective ways of depicting statistical information. Tufte has shown how we can design attractive graphs, maps and charts to help people to visualise data.[19] The best charts, he says, can be 'intriguing and curiosity-provoking'. Frugality and clarity are among his guiding principles. He warns against diagrams that are excessively elaborate, because the person creating them is attempting to impress the reader or demonstrate their artistic skills rather than focusing on the key message of the data. Three-dimensional bar charts, diagrams that use more ink than they need, and graphs with insufficient labels can dazzle and distort rather than inform. Tufte refers to many such efforts as 'chartjunk'.

Hans Rosling, who was renowned for swallowing a sword during his talks to prove that 'the seemingly impossible was possible', developed innovative ways of displaying statistics about economic development and health around the world.[20] These included animated graphics. On the gapminder.org website, which he founded, you can watch a kaleidoscope of balloons of different sizes and colours moving across a graph to display how life expectancy and income per person in different countries have increased together since 1800 (the sizes of the balloons represent the populations of the countries). Other dynamic maps and charts show how the populations and age structures of countries have changed dramatically over these years. And there are photographs of families around the world showing their weekly incomes in dollars, ranging from $29 for a family in Burundi to over $10,000 for a rich family in China.[21] You can read about each family and their lives and hopes for the future, and even look inside their homes to see the utensils they use and the toys their children play with. With displays like this, the families are no longer anonymous statistics belonging to far-off countries. Nor are they the stereotypes that we might expect; instead they show there is a diversity of wealth and lifestyles in each country.

There are many other examples of imaginative and effective ways of bringing numbers to life. The Brazilian scientist and designer Fernanda Bertini Viégas has created a wind map that you can see at: http://hint.fm/wind/. This displays continuously updated forecasts of wind speeds and directions across the USA on an animated map. In his book *Daily Rituals*, Mason Currey has a coloured chart that compares the ways that famous creative people apportioned their days and nights between sleeping, eating, exercising, relaxing, carrying out jobs, and engaging in creative activities.[22] It shows that Honoré de Balzac slept from 6.00 p.m. to 1.00 a.m. and spent the rest of the night on creative work, though he did have another sleep between 8.00 a.m. and 9.30 a.m. before resuming his writing. In contrast, Pablo Picasso and William Styron didn't usually rise until it was nearly lunchtime. Then there are the simple, but easily understood, charts created by the Open Data Institute to show trends in the British diet from 1974 to 2014, based on food diaries kept by families (you can find these at: http://britains-diet.labs.theodi.org/). They show that consumption of sugar, flour, potatoes and beverages have decreased over these years, while fruit, cereals and condiments have generally shown marked increases.

How many Eiffel Towers?

It's not always possible to turn numbers into pictures or charts – for example, they can take up a lot of space in newspapers or magazines. However, even written statements containing numbers can be made less numbing with a little care. Most of us have problems absorbing very large numbers, but they can often be cut down to a manageable size with a bit of thought. For example, it's been estimated that British women, aged 16 or over, spent £28.4 billion on clothes in 2017.[23] This doesn't mean much to me. But, given that there were roughly 27 million females in the UK who were aged 16 or over in 2017, that works out at about £1050 per woman. That number is easier to assimilate. Of course, we should remember that it's an average (mean) and so will probably have been pulled upwards by a few very high spenders.

The exact division of £28.4 billion by 27 million gives £1051.851852. But not only is expressing this number to six decimal places just as numbing as the figure expressed in billions, it's also spuriously accurate, since the numbers we used in the division were only rough estimates. Sometimes people like to create a false veneer of scientific exactitude by presenting numbers in this way. Dr Johnson said that round numbers are always false. But exact numbers are often false too, and even if they are not, it's frequently worth losing a bit of accuracy by rounding a number to get a message across that it is easily digested.

Another way of helping people to conceive large numbers when something is happening repeatedly is to use the 'rate per time period' method. For example, Greenpeace reported in April 2018 that the equivalent of one truckload of plastic enters the sea every minute of every day.[24] And, at one point in 2018, it was estimated that Jeff Bezos, the founder of Amazon, was earning nearly $2700 a second.[25]

Numbers can also make more sense if they are not presented in isolation. Comparisons and benchmarks can give them impact. I was once told that a typical bowl of porridge contains 3 to 4 grams of fibre, but is that good? I had a better idea when I found that this is equivalent to the fibre found in more than ten bowls of cornflakes. That phrase 'is equivalent to' can work wonders in turning abstract numbers into concrete images. USS *George H.W. Bush*, the giant American aircraft carrier, is 1092 feet long. I have difficulty imagining what this looks like, but that old journalistic standby – the length of a football pitch – can make the number easier to envisage. A typical football pitch (they can, of course, vary) is 345 feet long, so the length of the ship is equivalent to just over three football pitches.

For heights, the Empire State Building, the Statue of Liberty, the Shard in London, and the Eiffel Tower are frequently used comparators. In 2018, the Statue of Unity in Gujarat, India, was the tallest statue in the world. At 597 feet, the stunning bronze-clad sculpture of the Indian statesman Vallabhbhai Patel dwarfs the surrounding landscape like a scene from *Gulliver's Travels*. You can get a feel for what 597 feet looks like when you compare it to the Statue of Liberty – it's nearly twice as tall.

Words and numbers

When making comparisons, there is evidence from psychologists that we will make more use of numbers if they also come with labels telling us how good or bad they are. For example, when we make choices between different products or services, statistical information telling us about their performance on a range of attributes – such as the memory capacity, energy consumption and processor speed of laptop computers – might pass us by like ships in the night. But, if these are also labelled as 'poor', 'fair', 'good' or 'excellent', we tend to engage with the numeric information more.[26]

A combination of numbers and labels was also found to be most effective in conveying the uncertainty associated with temperature forecasts produced by the Intergovernmental Panel on Climate Change (IPCC).[27] For example, people made consistent and accurate interpretations of phrases such as 'very likely, that is, having a probability of greater than 90 per cent'. Of course, labels may reflect categories that have arbitrary boundaries, but unlike many of the examples we saw in Chapter 4, they are not replacing the numbers. Instead, they are encouraging us to make greater use of them.

A tyranny of numbers?

In the right hands, and with the right motives, numbers can be our allies. They can be attractive, engaging and informative, and their exactness and unambiguity can tell us things that words cannot. They can motivate us to achieve new goals and warn us of problems of which we might otherwise be unaware. They can reduce uncertainty and help us to clarify our thoughts so that our decisions can be made with greater insight and confidence. And they can challenge us when our views are ill-founded or misinformed.

But there is another side to the way we relate to numbers. A few years ago I read a book called *The Tyranny of Numbers* by David Boyle.[28] As the title implies, our association with numbers can sometimes make us feel like the passive numbed subjects of an authoritarian regime. It's a regime where numbers control our actions, determine our goals, establish our status, tell us self-serving falsehoods, and hide inconvenient truths. It is also a philistine regime that has no regard for art, beauty, love or spirituality. In this land, numbers are the property of a ruling intellectual elite, which denies its denizens the right to pry or question what they are saying. Obscure, and sometimes masked by technical terminology, the elite's metrics can seem to have little to do with people's day-to-day lives.

This book is intended to help people to put numbers back into their rightful place so that they become our honest servants, not our rulers. After all, we invented them. To achieve that, we need to be ready and able to challenge the inappropriate use of numbers. We need to be aware of their limitations while recognising and exploiting their strengths.

Ironically, we live in an era where burgeoning data sets, astonishing scientific advances, sophisticated methods of statistical analysis, lightning-speed computer processing power, and instant communications are juxtaposed with 'fake news' and a contempt for truthfulness by many key players in our societies. But, where numbers are concerned, many untruths are not deliberate. They can arise from well-intentioned people striving to measure elusive or immeasurable phenomena, or from oversimplifications designed to ease the communication of complex facts. They can trade on our yearning for simplicity, order and certainty where there is none. And, once accepted, they can be perpetuated because of our natural resistance to disconfirming facts.

However they arise, untruths are bad for our personal decisions and for democracy, so it pays to develop a healthy scepticism when a seductive number comes our way: to stand back, take a breath, and question how it was derived and what it truly represents – to think statistically. Indeed, it's apt to recall H.G. Wells's prediction made over a century ago, the one that helped me to get my first job: 'Statistical thinking will one day be as necessary for efficient citizenship as the ability to read and write.'

In a so-called post-truth age, that day has surely arrived.

Appendix

Questions to Ask about a Statistic

What's the motivation?

Ask yourself: what argument, product or service is the provider of a statistic trying to sell me? Numbers have a happy habit of supporting the views and interests of the person or organisation supplying them: '90 per cent fat free' to a food manufacturer is '10 per cent fat' to a nutritionist, even if the numbers are honest. These days there's no guarantee that numbers from politicians like Donald Trump will be truthful, but as individuals, we usually don't have the time or resources to check so-called facts for ourselves. If you have doubts, independent sites like the UK Statistics Authority, Full Fact, Channel 4 News FactCheck, and PolitiFact are well worth visiting.

What doesn't the number tell us?

Ask yourself, what does the number leave out? The punctuality of trains is not the same as the quality of service offered by the railway company. What about their trains' cleanliness and frequency, the friendliness of their staff, and the chances you'll be able to grab a seat or a cup of tea on board? Your resting heart rate is not the same as your cardiovascular health, and the number of exam passes by students at a school is not the same as the quality of education they have received. All of these are useful measures, but they only give us part of a more general picture. In particular, numbers tend to reflect the most easily quantifiable aspects of an issue – as we saw with GDP – and they can ignore qualitative factors completely. Many psychologists argue, for example, that IQ does not take into account a person's creativity. And, if you are told that your lifestyle increases your risk of suffering from a deadly disease by 30 per cent, without a starting figure the true risk you face may still be very small.

What simplifying assumptions underlie the number?

It's also worth asking in what ways the number simplifies the true picture and what assumptions it is based on. An average is a simplified summary of a diverse population and may not represent anyone at all. League table positions of cities are often based on weighted combinations of different attributes – such as pollution levels, the cost of living, and the frequency of crime, but the weights that are supposed to reflect the relative importance of these attributes may have absolutely no underlying rationale.

Then there are those arbitrary categories that can take on a life of their own once the dividing lines have been drawn that split numbers into groups – degree classifications, pupils' percentage marks transformed into letter grades, and weight classifications such as 'obese', 'overweight' and 'underweight'. They can make the world seem simpler, but the positions of the thresholds that divide the groups are often there for convenience, and like the weights in league tables, they don't always make sense. It's therefore worth avoiding the trap of assuming that membership of one category means that someone is very different to a member of another.

Is the number based on an unrepresentative survey?

An astonishing statistic in a newspaper might boost its circulation, but it's worth asking where the number came from. If it's based on a survey, was an effort made to question a representative cross section of the population that it claims to represent? People who are paid to be panellists in internet surveys may not be representative. Quota samples do try to ensure that the full diversity of the population is represented in the survey, but if you're never at home when the interviewer knocks on your door then you'll have no chance of being in the poll. People who tend to be at home, or on the street, at the times when the survey is being carried out may be unrepresentative of the general population. Even random samples can suffer from non-response from the people selected for interview. All of this means that published margins of error are likely to overestimate the reliability of a survey's findings.

Is the number based on a small sample?

Even surveys that are designed to be representative can produce inaccurate findings when only a small sample of people (or objects) has been used. Small samples can produce freak results. They can also fail to pick up on things that matter. For example, saying that a survey revealed no evidence that teaching method X is better than the current method being used might simply mean that too few children were studied to detect the benefits of method X. Be sceptical of findings from a survey if the sample size is not stated.

If a questionnaire was used to obtain the number, was it biased?

People's responses to a questionnaire can be manipulated by the way the questions are phrased. In particular, beware of statistics based on leading questions (Do you agree that this incompetent party should be voted out of office?), double-barrelled questions, where people might be saying yes to two things at once when they only agree with one of them (Do you agree that the government should spend more on defence and education?), presuming questions (When did you stop smoking?), and questions which unrealistically limit people's response options (Will you vote Labour or Conservative in the next election?).

Is the number based on subjective judgement?

If it is, we should be aware that people can have problems in converting their feelings into numbers, that psychological biases may distort how they respond, and that these responses can often be inconsistent. It's therefore unwise to treat subjective numbers as exact quantities. However, sometimes it's possible to say: 'Even if this subjective estimate changed by a considerable amount, we would still choose the same course of action, so it's safe to base our decision on it.'

Are any comparisons that are presented valid?

Not comparing like with like is an oft-rehearsed trick of those seeking to deceive us. Beware comparisons between two points in time that have been carefully selected to suggest an upward or downward trend where there is really no trend. Be especially wary of international comparisons where statistics have been collected using different definitions. Similarly, comparisons made in a single country over time can be invalid when the definition or scope of what is being measured has changed. In particular, distrust any comparisons of money spent or received that ignore the effects of price inflation.

Total or per capita?

If a country is spending 20 per cent more on healthcare than it was twenty years ago it might give us a warm feeling, but if the size of the population has dramatically increased over the same period then the expenditure per person might actually have gone down.

How reliable is the presentation of the number?

Look carefully at the vertical scales of graphs. They can sometimes be distorted or non-existent. Scales that don't start at zero can make small insignificant changes look large. Pictograms can also be used to exaggerate differences. For example, if the profits of a company have doubled, this could be depicted by presenting two dollar notes side by side with one twice as wide as the other. The problem is that to keep the shape of the larger dollar you would also have to double its length, making it its companion, so the increase in profits now looks much better than it really is. In graphs and tables, an 'Other' category can be used to bury important information that the supplier of the data does not want you to see.

Do the numbers make sense?

Finally, ask yourself if the numbers make sense. Analysts have found that hurricanes with female names cause more damage and destruction, more intelligent people prefer curly fries, and vegetarians miss fewer flights.[1] We saw earlier that our intuition is not a perfect guide on what we should believe or dismiss, so it's always worth asking: is there a plausible rationale to support the finding? If not, it's best to keep an open mind until further evidence is available.

Acknowledgements

The book would not have been possible without the support and encouragement of my wife, Chris. I am also most grateful to my agent, Peter Buckman of the Ampersand Agency, who provided expert advice, and to my editor at Profile Books, Ed Lake, whose detailed and insightful comments and suggestions enabled me to make considerable improvements to the draft manuscript. Eddie Mizzi's excellent copyediting led to further enhancements while Penny Daniel and her colleagues at Profile skillfully steered the book through the publication process.

I must also thank Professor Nick Kinnie for pointing me towards several useful sources of information. Over the years, many other colleagues and students have provided me with stories about the misuse and misinterpretation of statistics and also of examples of where they have led to important findings. Several of these accounts can be found in the book and I am indebted to those who alerted me to them.

Notes

Introduction

1. In *The Black Swan* (London: Allen Lane), Nassim Taleb refers to our tendency to produce explanations for links between events that are, in reality, unrelated as the narrative fallacy.
2. 'A comparison of European, American and Japanese values', Gallup Report, 1981.
3. http://www.slate.com/articles/news_and_politics/politics/2016/02/trump_is_winning_the_guy_you_d_want_to_have_a_beer_with_election.html.
4. Botsman, R. (2017). *Who Can You Trust? How Technology Brought Us Together – and Why It Could Drive us Apart*. London: Portfolio Penguin.
5. *The Week*, 3 June 2017.
6. Gronow, J. (2011). In: *Encylopedia of Consumer Culture* (ed. Dale Southerton). London: Sage, p. 256.
7. http://news.bbc.co.uk/1/hi/uk/302607.stm.
8. Savage, M. (2015). *Social Class in the 21st Century*. London: Pelican.
9. Lorne Jaffe, 'Five reasons why Facebook can be dangerous for people with depression', *Huffington Post* Blog, 15 May 2016.
10. https://www.washingtonpost.com/graphics/politics/trump-claims-database/ (updated to 13 November 2017).
11. Patrick Scott, *Daily Telegraph*, 8 March 2017.
12. *Daily Express*, 14 May 2016.
13. Kahneman, D. (2011). *Thinking Fast and Slow*. London: Allen Lane.

1. Rank Obsession

1. Josie Ensor, *The Guardian*, 7 July 2017.

2. *Allotment Wars*, broadcast on BBC One on 22 January 2013.
3. Anthony Faiola, *The Washington Post*, 14 December 2013.
4. https://www.statista.com/statistics/248335/number-of-new-titles-and-re-editions-in-selected-countries-worldwide/.
5. Tversky, A. (1969). 'Intransitivity of preferences', *Psychological Review*, 76, 31–48.
6. Tversky, A. (1972). 'Elimination by aspects a theory of choice', *Psychological Review*, 79, 281–299.
7. Gigerenzer, G., Todd, P.M, & the ABC Research Group (1999). *Simple Heuristics that Make Us Smart*. Oxford University Press.
8. Arrow, K. (1950). 'A difficulty in the concept of social welfare', *Journal of Political Economy*, 58, 328–346.
9. BBC News, 'Labour suffers but may hold on', 4 May 2007.
10. Janis, I.R. (1982). *Groupthink*, 2nd edn. Boston, MA: Houghton Mifflin.
11. Rank reversal can occur for a number of reasons. The following simple example illustrates one of the causes. Two universities, A and B, are to be ranked on two criteria: (i) the average number of research papers published by members of staff in the last two years, and (ii) the amount each university spends per student on teaching and library facilities. The table below shows these figures. The person compiling the rankings judges that a weight of 70 per cent should be assigned to the number of research papers published and 30 per cent to the expenditure per student. He thinks this reflects the relative importance of the criteria, but notice that the number of papers published per staff member at university B is very close to that of A.

University	Papers published per staff member	Expenditure per student ($)
A	10	2000
B	9	4000
Weight	70%	30%

Because the two criteria are expressed in different units these are each converted to a score on 0 (worst performance) to 100 (best performance) scales. This makes them easier to compare and is common practice when league tables are constructed.

University	Papers published per staff member	Expenditure per student
A	100	0
B	0	100
Weight	70%	30%

We can now obtain an overall score for each university by multiplying each weight by the appropriate score and adding the results. For university A we have (70% × 100) + (30% × 0) = 70. A similar calculation yields a score of 30 for B. So A beats B to the number one position.

Let's now add a third university, C, to the rankings. The table below shows its performance alongside those of A and B.

University	Papers published per staff member	Expenditure per student ($)
A	10	2000
B	9	4000
C	5	5000
Weights	70%	30%

When we convert these measures to 0 to 100 scales we obtain the following table. For example, the number of papers published per member of staff at university B is 4/5th of the way between C and A so it now gets a score of 80.

University	Papers published per staff member	Expenditure per student
A	100	0
B	80	67
C	0	100
Weights	70%	30%

Multiplying the weights by the scores for each university and adding the results gives the following overall scores. University A: 70, university B: 76, university C: 30. So now university B beats A and achieves the top rank.

What has happened? When there are just two universities the large weight of 70 per cent assigned to the number of papers published per staff member ignored the fact that there was only a small difference between A and B's performances. This meant that B's inferiority to A on that measure was exaggerated. However, when C was added it lagged a long way behind the other universities on the number of papers published so the range of performances on this criterion was greater. This boosted B's score on the 0 to 100 scale – it went from 0 to 80. When combined with the weight of 70 per cent, this was sufficient for it to overtake A's overall score.

The key point is that the 0 to 100 scales disguise the fact that the differences between the best and worst performances on some criteria can be relatively unimportant. Generally, a bigger difference should be associated with a larger weight. An improvement in the average number of papers published from 5 to 10 is more important than a small increase from 9 to 10. If, as in this example, the weights don't reflect these differences then anomalies like rank reversal are possible.

For other explanations of the phenomenon, see: Tofallis, C. (2014). 'Add or multiply? A tutorial on ranking and choosing with multiple criteria', *INFORMS Transactions on Education*, 14, 109–119.

12. Von Winterfeldt, D., & Edwards, W. (1986). *Decision Analysis and Behavioural Research*. Cambridge University Press.
13. Broede Carmody & Aisha Dow, 'Top of the world: Melbourne crowned world's most liveable city, again', *The Age*, 18 August 2016.
14. 'How can you build a strong city pulse, without taking the human pulse?', Ernst and Young Report, July 2016.
15. Broede Carmody & Aisha Dow, 'Top of the world: Melbourne crowned world's most liveable city, again', *The Age*, 18 August 2016.
16. See: Brendan F.D. Barrett, https://ourworld.unu.edu/en/the-worlds-most-liveable-cities.
17. Kahneman, D., & Tversky, A. (1979). 'Prospect theory: An analysis of decision under risk', *Econometrica*, 47, 263–291.
18. AARP Liveability Index.
19. Boyle, D. (2000). *The Tyranny of Numbers*. London: HarperCollins.
20. http://www.goodnet.org/articles/meet-city-that-measures-smiles-per-hour.
21. Boyle, D. (2000). *The Tyranny of Numbers*. London: HarperCollins.
22. Quoted in: Derksen, W. (2017). *The Way of Letting Go: One Woman's Walk toward Forgiveness*. London: HarperCollins.
23. Anderson, C., Hildreth, J.A.D., & Howland, L. (2015). 'Is the desire

for status a fundamental human motive? A review of the empirical literature', *Psychological Bulletin*, 141, 574–601.
24. Editorial, *The Guardian*, 1 April 2019.

2. Perilous Proxies

1. Chatterjee, A., & Hambrick, D.C. (2007). 'It's all about me: Narcissistic chief executive officers and their effects on company strategy and performance', *Administrative Science Quarterly*, 52, 351–386.
2. Emmons, R. (1987). 'Narcissism: Theory and measurement', *Journal of Personality and Social Psychology*, 52, 11–17.
3. Brookers, M. (2004). *Extreme Measures: The Dark Visions and Bright Ideas of Francis Galton*. New York: Bloomsbury.
4. Ibid.
5. https://urbanisation.econ.ox.ac.uk/blog/nighttime-lights-how-are-they-useful-dzhamilya-nigmatulina.
6. Kounali, D., Robinson, T., Goldstein, H., & Lauder, H. (2008). 'The probity of free school meals as a proxy measure for disadvantage', Working paper, University of Bath.
7. Der, G., & Deary, I.J. (2017). 'The relationship between intelligence and reaction time varies with age: Results from three representative narrow-age age cohorts at 30, 50 and 69 years, *Intelligence*, 64, 89–97. See also: Johnson, R.C., McClearn, G.E., Yuen, S., Nagoshi, C.T., Ahern, F.M., & Cole, R.E. (1985). 'Galton's data a century later', *American Psychologist*, 40, 875–892.
8. Fan, M.D.M. (2007). 'The immigration–terrorism illusory correlation and heuristic mistake', *Harvard Latino Law Review*, 10, 33–52.
9. https://repositori.upf.edu/bitstream/handle/10230/4573/1132.pdf.
10. Gregory, R., Failing, L., Harstone, M., Long, G., & McDaniels, T. (2012). *Structured Decision Making: A Practical Guide to Environmental Management*. Hoboken, NJ: Wiley.
11. Murray, D., Schwartz, J.B., & Lichter, R.S. (2001). *It Ain't Necessarily So: How Media Make and Unmake the Scientific Picture of Reality*. Lanham, MD: Rowman & Littlefield, pp. 75–76.
12. See, for example: https://www.theguardian.com/society/2010/feb/24/mid-staffordshire-hospital-inquiry.
13. https://www.kingsfund.org.uk/projects/general-election-2010/performance-targets.
14. Reynaert, M., & Sallee, J.M. (2016). 'Corrective policy and Goodhart's

law: The case of carbon emissions from automobiles', National Bureau of Economic Research Working Paper 22911.
15. https://www.independent.co.uk/news/business/news/budget-2018-philip-hammond-analysis-what-it-means-growth-austerity-spending-tax-obr-brexit-a8607661.html.
16. Sarah O'Connor, *The Financial Times*, 29 May 2014.
17. https://www.bbc.co.uk/news/business-26913497.
18. See: Coyle, D. (2014). *GDP: A Brief but Affectionate History*. Woodstock, Oxon: Princeton University Press, p. 114.
19. Papanicolas, I., Woskie, L.R., & Jha, A.K. (2018). 'Health care spending in the United States and other high-income countries', *JAMA*, 319, 1024–1039.
20. Aichner, T., & Coletti, P. (2013). 'Customers' online shopping preferences in mass customization', *Journal of Direct Data and Digital Marketing Practice*, 15, 20–35.
21. See: Coyle, D. (2014). *GDP: A Brief but Affectionate History*. Woodstock, Oxon: Princeton University Press, pp. 122–124.
22. Ibid., p. 127.
23. Tansy Hoskins, 'Cotton production linked to images of the dried up Aral Sea basin', *The Guardian*, 1 October 2014.
24. Rebecca Smithers, 'UK households binned 300,000 tonnes of clothing in 2016', *The Guardian*, 11 July 2017.
25. https://ec.europa.eu/eurostat/documents/118025/118123/Fitoussi+Commission+report.
26. https://www.independent.co.uk/news/uk/home-news/violent-crime-sex-offences-railways-trains-british-transport-police-figures-a8569486.html.
27. http://media.btp.police.uk/r/15934/british_transport_police_releases_its_annual_repo.
28. Wolfgang, M., Figlio, R.M., Tracy, P.E. & Singer, S.I. (1985). *The National Survey of Crime Severity*. Washington, DC: US Department of Justice, Bureau of Justice Statistics.
29. Sherman, L., Neyroud, P.W., & Neyroud, E. (2016). 'The Cambridge Crime Harm Index: Measuring total harm from crime based on sentencing guidelines', *Policing: A Journal of Policy and Practice*, 10, 171–183.
30. Ashby, M.P. (2017). 'Comparing methods for measuring crime harm/severity', *Policing: A Journal of Policy and Practice* 12, 439–454.

31. https://www.sentencingcouncil.org.uk/offences/crown-court/item/preparation-of-terrorist-acts/.
32. Note that crimes recorded arising from police action that set out to detect hitherto unreported crimes – such as drug searches or arrests of motorists during police monitoring of traffic – are not counted in the index. These crimes reflect levels of police activity and will increase or decrease depending on the intensity of that activity.
33. Ashby, M.P. 'Comparing methods for measuring crime harm/severity', *Policing: A Journal of Policy and Practice* 12, 439–454.
34. Moore, R.H. (1984). 'Shoplifting in middle America: Patterns and motivational correlates', *International Journal of Offender Therapy and Comparative Criminology*, 28(1), 53–64.
35. Sherman, L., Neyroud, P.W., & Neyroud, E. (2016). 'The Cambridge Crime Harm Index: Measuring total harm from crime based on sentencing guidelines', *Policing: A Journal of Policy and Practice*, 10, 171–183.
36. Danziger, S., Levav, J., & Avnaim-Pesso, L. (2011). 'Extraneous factors in judicial decisions', *Proceedings of the National Academy of Sciences*, 108, 6889–6892.
37. Ashby, M.P. 'Comparing methods for measuring crime harm/severity', *Policing: A Journal of Policy and Practice* 12, 439–454.
38. Doob, A.N., & Gross, A.E. (1968). 'Status of frustrator as an inhibitor of horn-honking responses', *The Journal of Social Psychology*, 76, 213–218.
39. Hubbard, D.W. (2010). *How to Measure Anything*, 2nd edition. Hoboken, NJ: Wiley.
40. http://sinsofgreenwashing.com/index6b90.pdf.
41. 'Ten worst household products for greenwashing', CBC News Canada, 14 September 2012.

3. One Number Says It All

1. Mona Chalabi, *The Guardian*, 28 November, 2013.
2. Ritchie, S. (2015). *Intelligence: All That Matters*. London: Hodder & Stoughton.
3. Antonios, N., & Raup, C., (2012). 'Buck v. Bell (1927)', *Embryo Project Encyclopedia*: http://embryo.asu.edu/handle/10776/2092.
4. www.globalresearch.ca/us-court-ruled-you-can-be-too-smart-to-be-a-cop/5420630.
5. Smith, C.R. (2004). *Learning Disabilities: The Interaction of Students and Their Environments*. Boston, MA: Allyn & Bacon.

6. Ritchie, S. (2015). *Intelligence: All That Matters*. London: Hodder & Stoughton.
7. Gottfredson, L.S. (1997). 'Mainstream science on intelligence', *Intelligence*, 24, 13–23.
8. Daphne Martschenko, 'The IQ test wars: Why screening for intelligence is still so controversial', *The Conversation*, 24 January 2018.
9. Scott Barry Kaufman, 'Intelligent testing: The evolving landscape of IQ testing', *Psychology Today*, https://www.psychologytoday.com/gb/blog/beautiful-minds/200910/intelligent-testing.
10. Schneider, W.J., & Newman, D.A. (2015). 'Intelligence is multidimensional: Theoretical review and implications of specific cognitive abilities', *Human Resource Management Review*, 25, 12–27.
11. https://www.prnewswire.com/news-releases/reebok-survey-humans-spend-less-than-one-per cent-of-life-on-physical-fitness-300261752.html.
12. Source: www.disabled-world.com.
13. www.richardwiseman.com/quirkology/pace_home.htm.
14. Robinson, W.S. (2009). 'Ecological correlations and the behavior of individuals', *International Journal of Epidemiology*, 38, 337–341.
15. Gelman, A., Shor, B., Bafumi, J., & Park, D. (2007). 'Rich state, poor state, red state, blue state: What's the matter with Connecticut?', *Quarterly Journal of Political Science*, 2, 345–367.
16. For simplicity, assume that only five voters live in each state. The table below shows their annual income and the probability that they will vote Republican.

	State A		State B		State C	
	Income ($)	*Probability (%)*	*Income ($)*	*Probability (%)*	*Income ($)*	*Probability (%)*
Voter 1	20000	55	30000	35	30000	10
Voter 2	30000	60	40000	40	40000	20
Voter 3	40000	65	55000	50	60000	40
Voter 4	50000	70	60000	60	70000	50
Voter 5	60000	80	70000	75	80000	60
Mean	40000	66	51000	52	56000	36

Within each state, the more a voter earns, the higher is the probability that they will vote Republican. But, looking at the means, the higher the average income is in a state, then the lower is the average probability that people will vote Republican.

17. Some people dispute the claim that Durkheim was subject to the ecological fallacy. See: Berk, B.B. (2006). 'Macro–micro relationships in Durkheim's analysis of egoistic suicide', *Sociological Theory*, 24, 58–80.
18. Carroll, K. (1975). 'Experimental evidence of dietary factors and hormone-dependent cancers', *Cancer Research*, 35, 3374–3383.
19. Holmes, M.D., Hunter, D.J., Colditz, G.A., Stampfer, M.J., Hankinson, S.E., Speizer, F.E., Rosner, B., & Willett, W.C. (1999). 'Association of dietary intake of fat and fatty acids with risk of breast cancer', *Journal of the American Medical Association*, 281, 914–920.
20. Myers, D.G. (2000). 'The funds, friends, and faith of happy people', *American Psychologist*, 55, 56–67. See also: Dolan, P. (2014). *Happiness by Design*. London: Penguin (see Chapter 7 on how happiness is assessed), and Oliver Burkeman, *Guardian Weekly*, 18 January 2019.
21. Mancini, A.D. (2013). 'The trouble with averages: The impact of major life events and acute stress may not be what you think', Council on Contemporary Families Research Brief.

4. Leaps and Boundaries

1. *New York Times*, 13 July 2005.
2. Kevin McCoy, 'Merck to face first Vioxx trial before Texas jury next month', *USA Today*, 19 June 2005.
3. 'Merck CEO resigns as drug probe continues', *Washington Post*, 6 May 2005.
4. Technically, it's better to say there is insufficient evidence to reject the hypothesis, rather than saying it is accepted, hence the inverted commas.
5. Alex Berenson, 'Evidence in Vioxx suits shows intervention by Merck officials', *New York Times*, 24 April 2005.
6. Ziliak, S.T., & McCloskey, D.N. (2008). *The Cult of Statistical Significance: How the Standard Error Cost Us Jobs, Justice, and Lives*. Ann Arbor, MI: University of Michigan Press.
7. Alex Berenson, 'Evidence in Vioxx suits shows intervention by Merck officials', *New York Times*, 24 April 2005.
8. https://www.npr.org/templates/story/story.php?storyId=5470430.
9. 'When half a million Americans died and nobody noticed', *The Week*, 2 April 2012.
10. https://fivethirtyeight.com/features/science-isnt-broken/#part1.
11. Simmons, J.P., Nelson, L.D., & Simonsohn, U. (2011). 'False-positive

psychology: Undisclosed flexibility in data collection and analysis allows presenting anything as significant', *Psychological Science*, 22, 1359–1366.
12. Gigerenzer, G. (2004). 'Mindless statistics', *The Journal of Socio-Economics*, 33, 587–606.
13. Goodwin, P. (2007). 'Should we be using significance tests in forecasting research?', *International Journal of Forecasting*, 23, 333–334.
14. Sabrina Barr, 'Eating red meat and cheese can help heart health, scientists claim', *The Independent*, 31 August 2018.
15. Quotes taken from: Robert Matthews, 'Silly science', *Prospect*, November, 1998. Our focus here is on the arbitrary nature of the '$p < 0.05$' rule. There are many other problems with significance tests. See: Ziliak, S.T., & McCloskey, D.N. (2008). *The Cult of Statistical Significance: How the Standard Error Cost Us Jobs, Justice, and Lives*. Ann Arbor, MI: University of Michigan Press.
16. Greenland, S., Senn, S.J., Rothman, K.J., Carlin, J.B., Poole, C., Goodman, S.N., Altman, D.G. (2016). 'Statistical tests, P values, confidence intervals, and power: A guide to misinterpretations', *European Journal of Epidemiology*, 31, 337–350.
17. Wasserstein, R.L., & Lazar, N.A. (2016). 'The ASA's statement on *p*-values: Context, process, and purpose', *The American Statistician*, 70, 129–133.
18. Ioannidis, J.P. (2005). 'Why most published research findings are false', *PLoS Medicine*, 2(8), e124. See also: Leek, J.T., & Jager, L.R. (2017). 'Is most published research really false?', *Annual Review of Statistics and Its Application*, 4, 109–122.
19. Neville, P. (2013). *Historical Dictionary of British Foreign Policy*, Lanham, MD: Scarecrow Press. Some other sources indicate that he got a 'third'.
20. https://www.telegraph.co.uk/education/2017/08/24/passing-exam-has-never-easier-just-15-per-cent-required-pass/.
21. Rothbart, M., Davis-Stitt, C., & Hill, J. (1997). 'Effects of arbitrarily placed category boundaries on similarity judgements', *Journal of Experimental Social Psychology*, 33, 122–145.
22. Galak, J., Kruger, J., & Rozin, P. (2009). 'Not in my backyard: The influence of arbitrary boundaries on consumer choice', in: Ann L. McGill & Sharon Shavitt (eds), *NA – Advances in Consumer Research*, volume 36. Duluth, MN: Association for Consumer Research, pp. 79–81.
23. Flegal, K.M., Kit, B.K., & Graubard, B.I. (2014). 'Body mass index categories in observational studies of weight and risk of death', *American Journal of Epidemiology*, 180, 288–296.

24. Nuttall, F.Q. (2015). 'Body mass index. Obesity, BMI, and health: A critical review', *Nutrition Today*, 50, 117–128.
25. Foroni, F., & Rothbart, M. (2013). 'Abandoning a label doesn't make it disappear: The perseverance of labeling effects', *Journal of Experimental Social Psychology*, 49, 126–131.
26. Krueger, J., & Clement, R.W. (1994). 'Memory-based judgements about multiple categories: A revision and extension of Tajfel's accentuation theory. *Journal of Personality and Social Psychology*, 67, 35–47.
27. Rothbart, M., Davis-Stitt, C., & Hill, J. (1997). 'Effects of arbitrarily placed category boundaries on similarity judgments', *Journal of Experimental Social Psychology*, 33, 122–145.
28. Hunger, J.M., & Tomiyama, A.J. (2014). 'Weight labeling and obesity: A longitudinal study of girls aged 10 to 19 years', *JAMA Pediatrics*, 168, 579–580.
29. Foroni, F. (2005). 'Labeling and categorization: Evidence for a mere labeling effect, its modulating factors, and characteristics', Doctoral Dissertation, University of Oregon.
30. Rosenthal, R., & Jacobson, L. (1992). *Pygmalion in the Classroom: Teacher Expectation and Pupils' Intellectual Development*. New York: Irvington. See also: Adam Alter, 'Why it's dangerous to label people', *Psychology Today* Blog, 17 May 2010.
31. Freyd, J.J. (1983). 'Shareability: The social psychology of epistemology', *Cognitive Science*, 7, 191–210.

5. A Daily Life in Numbers

1. Scarlett Thomas, 'Nowhere to run: Did my fitness addiction make me ill?', *The Guardian*, 7 March 2015.
2. Lupton, D. (2015). 'Quantified sex: A critical analysis of sexual and reproductive self-tracking using apps', *Culture, Health & Sexuality*, 17, 440–453.
3. *Metro*, 25 January 2019.
4. Gary Wolf, 'Know thyself: Tracking every facet of life, from sleep to mood to pain, 24/7/365', *Wired*, 22 June, 2009.
5. Danaher, J., Nyholm, S., & Earp, B.D. (2018). 'The quantified relationship', *The American Journal of Bioethics*, 18, 3–19.
6. https://www.geekwire.com/2012/kouply-mobile-game-save-marriage/.
7. Lupton, D. (2016). *The Quantified Self*. Cambridge, UK: Polity Press, p. 46.
8. Ibid., pp. 76–77.
9. https://www.bbc.co.uk/programmes/b02mfzrw.

10. Rachel Bachman, 'Want to cheat your Fitbit? Try a puppy or a power drill', *Wall Street Journal*, 9 June 2016.
11. http://quantifiedself.com/2013/02/larry_smarr_croneshope_in_data/#more-5853.
12. Lupton, D. (2016). *The Quantified Self*. Cambridge, UK: Polity Press, p. 140.
13. Danaher, J., Nyholm, S., & Earp, B.D. (2018). 'The benefits and risks of quantified relationship technologies: Response to open peer commentaries on "The Quantified Relationship"', *The American Journal of Bioethics*, 18, W3–W6. The danger that apps might violate users' privacy is another concern. See, for example, Stuart Dredge, 'Yes, those free health apps are sharing your data with other companies', *The Guardian*, 3 September, 2013. The focus here, however, is on their ability to accurately reflect what they say they are measuring.
14. Reynolds, J.M. (2018). 'Infotality: On living, loving, and dying through information', *American Journal of Bioethics*, 18, 33–35.
15. https://www.medicalnewstoday.com/articles/283117.php.
16. https://www.jefftk.com/p/happiness-logging-one-year-in.
17. http://quantifiedself.com/2010/04/why-i-stopped-tracking/.
18. Morozov, E. (2013). *To Save Everything, Click Here: Technology, Solutionism, and the Urge to Fix Problems that Don't Exist*. London: Allen Lane.
19. Sharon, T., & Zandbergen, D. (2017). 'From data fetishism to quantifying selves: Self-tracking practices and the other values of data', *New Media & Society*, 19, 1695–1709.
20. Phillips, L.D. (1984). 'A theory of requisite decision models', *Acta Psychologica*, 56, 29–48.
21. Danaher, J., Nyholm, S., & Earp, B.D. (2018). 'The quantified relationship', *The American Journal of Bioethics*, 18, 3–19.

6. Polls Apart

1. https://fullfact.org/europe/factcheck-daily-express-eu-poll-biased-and-wide-mark/.
2. These headlines are reproduced at: https://www.buzzfeed.com/scottybryan/the-crazy-daily-express-99-polls-b7bm.
3. https://www.genealogybranches.com/censuscosts.html.
4. Kate Allen, 'Researchers in UK count cost of plan to scrap census', *Financial Times*, 1 September 2013.
5. Adam Corey Ross, 'Cost of 2010 Census a whopping $14.7 billion', *The Fiscal Times*, 13 January 2011.

6. Andrew McCorkell, 'Fears for next year's census after errors in 2001', *The Guardian*, 27 December 2010.
7. In random digit dialling (RDD), a computer randomly selects and dials telephone numbers in the hope that interviews can be conducted with whoever answers.
8. Britain's premium bond computer, ERNIE, an acronym for Electronic Random Number Identification Equipment, does produce true random numbers. The latest version uses quantum technology to produce random numbers from light.
9. Tom Cardoso, '"Anything would be better:" Critics warn Ottawa's family-reunification lottery is flawed, open to manipulation', *The Globe and Mail*, 8 June 2019.
10. There's also an ironic problem for election pollsters in countries like Britain where voting is not compulsory. It seems sensible to strive for a representative cross section of adults. But the population of adults is not the same as the population of people who actually turn out to vote. People might tell the pollsters they'll vote and then decide to stay at home on election day. Only 66 per cent of eligible voters took part in Britain's 2015 General Election and these tended to be older, more affluent, and better educated. The result was a shock result on election night both for the pollsters and politicians. See: https://www.britishelectionstudy.com/bes-resources/why-the-polls-got-it-wrong-and-the-british-election-study-face-to-face-survey-got-it-almost-right/#.XP1W5Obruit.
11. *The Superpollsters: How They Measure and Manipulate Public Opinion in America*. New York: Four Walls Eight Windows. See also: Warren, K.F. (2018). *In Defense of Public Opinion Polling*. New York: Taylor and Francis.
12. Fricker, R.D. (2017). 'Sampling methods for online surveys,' in: N.G. Fielding, R.M., Lee, and G. Blank (eds), *The Sage Handbook of Online Research Methods*, 2nd edition. London: Sage.
13. Kennedy, C., Mercer, A., Keeter, S., Hatley, N., McGeeney, K., and Gimenez, A. (2016). 'Evaluating online nonprobability surveys', Pew Research Center Report.
14. Hillygus, D.S., Jackson, N., and Young, M. (2014), 'Professional respondents in nonprobability online panels', in: M. Callegaro, R. Baker, J. Bethlehem, A.S. Göritz, J.A. Krosnick, and P.J. Lavrakas (eds.), *Online Panel Research: A Data Quality Perspective*. Chichester: Wiley, pp. 219–237.
15. Prosser, C., & Mellon, J. (2018). 'The twilight of the polls? A review of

trends in polling accuracy and the causes of polling misses', *Government and Opposition*, 53, 757–790.

16. Keeter, S., Hatley, N., Kennedy, C., and Lau, A. (2017). 'What low response rates mean for telephone surveys', Pew Research Center Report.
17. Rivers, D. (2013). 'Comment on Task Force Report', *Journal of Survey Statistics and Methodology*, 1, 111–17.
18. Tim Marcin, 'Support for Donald Trump's impeachment is higher than his approval rating, new poll shows', *Newsweek*, 22 January 2019.
19. Peterson, R.A. (2018). 'On the myth of reported precision in public opinion polls', *International Journal of Market Research*, 60, 147–155.
20. https://www.newscientist.com/article/2203837-how-did-pollsters-get-the-australian-election-result-so-wrong/.
21. Bhatti, Y., & Pedersen, R.T. (2015). 'News reporting of opinion polls: Journalism and statistical noise', *International Journal of Public Opinion Research*, 28, 129–141. See also: Tryggvason Oleskog, P., & Strömbäck, J. (2018). 'Fact or fiction? Investigating the quality of opinion poll coverage and its antecedents', *Journalism Studies*, 19, 2148–2167.
22. Alima Hotakie, 'How big a role does luck play in football?', *Aljazeera News*, 13 July 2018.
23. Du, N., Budescu, D.V., Shelly, M.K., & Omer, T.C. (2011). 'The appeal of vague financial forecasts', *Organizational Behavior and Human Decision Processes*, 114(2), 179–189. See also: Gaertig, C., & Simmons, J.P. (2018). 'Do people inherently dislike uncertain advice?', *Psychological Science*, 29, 504–520.
24. Yaniv, I., & Foster, D.P. (1995). 'Graininess of judgement under uncertainty: An accuracy–informativeness trade-off', *Journal of Experimental Psychology: General*, 124, 424–432.
25. Toff, B. (2018). 'Exploring the effects of polls on public opinion: How and when media reports of policy preferences can become self-fulfilling prophesies', *Research & Politics*, 5, 1–9.
26. McAllister, I., and Studlar, D.T. (1991) 'Bandwagon, underdog, or projection? Opinion polls and electoral choice in Britain, 1979–1987', *The Journal of Politics*, 53, 720–741. See also: Dahlgaard, J.O., Hansen, J.H., Hansen, K.M., & Larsen, M.V. (2017). 'How election polls shape voting behaviour', *Scandinavian Political Studies*, 40, 330–343.
27. Rogers, T., and Moore, D.A. (2014). 'The motivating power of under-confidence: "The race is close but we're losing"', HKS Working Paper No. RWP14–047.

28. Jennings, W., & Wlezien, C. (2018). 'Election polling errors across time and space', *Nature Human Behaviour*, 2, 276–283.
29. Sarah Marsh, 'Britons get drunk more often than 35 other nations, survey finds', *The Guardian*, 15 May 2019.

7. On a Scale of 1 to 10, How Do You Feel?

1. Jahedi, S., & Méndez, F. (2014). 'On the advantages and disadvantages of subjective measures', *Journal of Economic Behavior & Organization*, 98, 97–114.
2. As reported in: Packard, Vance (1957). *The Hidden Persuaders*. London: Longmans, Green, and Co.
3. Bertrand, M., & Mullainathan, S. (2001). 'Do people mean what they say? Implications for subjective survey data', *American Economic Review*, 91, 67–72. Silver, B.D., Anderson, B.A., & Abramson, P.R. (1986). 'Who overreports voting?', *American Political Science Review*, 80, 613–624.
4. Fisher, T.D. (2013). 'Gender roles and pressure to be truthful: The bogus pipeline modifies gender differences in sexual but not non-sexual behavior', *Sex Roles*, 68, 401–414.
5. Livingston, M., & Callinan, S. (2015). 'Underreporting in alcohol surveys: Whose drinking is underestimated?', *Journal of Studies on Alcohol and Drugs*, 76, 158–164.
6. Loureiro, M.L., & Lotade, J. (2005). 'Interviewer effects on the valuation of goods with ethical and environmental attributes', *Environmental and Resource Economics*, 30, 49–72.
7. Larson, R.B. (2018). 'Examining consumer attitudes toward genetically modified and organic foods', *British Food Journal*, 120, 999–1014.
8. Smith, B., Olaru, D., Jabeen, F., & Greaves, S. (2017). 'Electric vehicles adoption: Environmental enthusiast bias in discrete choice models', *Transportation Research Part D: Transport and Environment*, 51, 290–303.
9. Costanigro, M., McFadden, D.T., Kroll, S., & Nurse, G. (2011). 'An in-store valuation of local and organic apples: The role of social desirability', *Agribusiness*, 27, 465–477.
10. Klaiman, K., Ortega, D.L., & Garnache, C. (2016). 'Consumer preferences and demand for packaging material and recyclability', *Resources, Conservation and Recycling*, 115, 1–8.
11. Olynk, N.J., Tonsor, G.T., & Wolf, C.A. (2010). 'Consumer willingness to pay for livestock credence attribute claim verification', *Journal of Agricultural and Resource Economics*, 35(2), 261–280.
12. Miller, N.J., & Kean, R.C. (1997). 'Reciprocal exchange in rural

communities: Consumers' inducements to inshop', *Psychology & Marketing*, 14, 637–661.

13. Bateman, I.J., & Mawby, J. (2004). 'First impressions count: Interviewer appearance and information effects in stated preference studies', *Ecological Economics*, 49, 47–55.
14. Krosnick, J.A., & Alwin, D.F. (1987). 'An evaluation of a cognitive theory of response-order effects in survey measurement', *Public Opinion Quarterly*, 51(2), 201–219.
15. Bradburn, N.M., Rips, L.J., & Shevell, S.K. (1987). 'Answering autobiographical questions: The impact of memory and inference on surveys', *Science*, 236, 157–161.
16. Wagenaar, W.A. (1986). 'My memory: A study of autobiographical memory over six years', *Cognitive Psychology*, 18, 225–252.
17. Hartley, E. (1946). *Problems in Prejudice*. New York: Octagon Press.
18. Bishop, G.F., Tuchfarber, A.J., & Oldendick, R.W. (1986). 'Opinions on fictitious issues: The pressure to answer survey questions', *Public Opinion Quarterly*, 50, 240–250.
19. Sturgis, P., & Smith, P. (2010). 'Fictitious issues revisited: Political interest, knowledge and the generation of nonattitudes', *Political Studies*, 58, 66–84.
20. Zaller, J., & Feldman, S. (1992). 'A simple theory of the survey response: Answering questions versus revealing preferences', *American Journal of Political Science*, 579–616.
21. Meeran, S., Jahanbin, S., Goodwin, P., & Quariguasi Frota Neto, J. (2017). 'When do changes in consumer preferences make forecasts from choice-based conjoint models unreliable?', *European Journal of Operational Research*, 258, 512–524.
22. Zaller, J., & Feldman, S. (1992). 'A simple theory of the survey response: Answering questions versus revealing preferences', *American Journal of Political Science*, 36, 579–616.
23. Strack, F., Martin, L.L., & Schwarz, N. (1988). 'Priming and communication: Social determinants of information use in judgements of life satisfaction', *European Journal of Social Psychology*, 18, 429–442.
24. Van de Walle, S., & Van Ryzin, G.G. (2011). 'The order of questions in a survey on citizen satisfaction with public services: Lessons from a split-ballot experiment', *Public Administration*, 89, 1436–1450.
25. Schuldt, J.P., Konrath, S.H., & Schwarz, N. (2011). '"Global warming" or "climate change"? Whether the planet is warming depends on question wording', *Public Opinion Quarterly*, 75, 115–124.

26. Schwarz, N., Knäuper, B., Hippler, H.J., Noelle-Neumann, E., & Clark, L. (1991). 'Rating scales numeric values may change the meaning of scale labels', *Public Opinion Quarterly*, 55, 570–582.
27. Bjørnskov, C. (2010). 'How comparable are the Gallup World Poll life satisfaction data?', *Journal of Happiness Studies*, 11, 41–60.
28. http://worldhappiness.report/.
29. https://www.ons.gov.uk/peoplepopulationandcommunity/wellbeing/bulletins/measuringnationalwellbeing/april2017tomarch2018.
30. Schwarz, N. (2011). 'Feelings-as-information theory', in: Van Lange, P., Kruglanski, A., & Higgins, E.T. (eds), *Handbook of Theories of Social Psychology*, Vol. 1, 289–308. London: Sage.
31. Song, H., & Schwarz, N. (2009). 'If it's difficult to pronounce, it must be risky: Fluency, familiarity, and risk perception', *Psychological Science*, 20, 135–138.
32. Bjørnskov, C. (2010). 'How comparable are the Gallup World Poll life satisfaction data?', *Journal of Happiness Studies*, 11, 41–60.
33. Kahneman, D., & Tversky, A. (2013). 'Prospect theory: An analysis of decision under risk', In *Handbook of the Fundamentals of Financial Decision Making: Part I*, pp. 99–127.
34. Kahneman, D. (2011). *Thinking Fast and Slow*. London: Allen Lane, p. 405.
35. Susanna Rustin, 'Can happiness be measured?', *The Guardian*, 20 July 2012.
36. Mark Holder, 'Measuring happiness: How can we measure it?', *Psychology Today*, 22 May 2017: https://www.psychologytoday.com/gb/blog/the-happiness-doctor/201705/measuring-happiness-how-can-we-measure-it.
37. Julian Baggini, 'Why it's impossible to measure happiness', *Prospect*, 18 October 2018.
38. Kahneman, D. (2011). *Thinking Fast and Slow*. London: Allen Lane, p. 405.
39. Hardy, J.D., & Javert, C.T. (1949). 'Studies on pain: Measurements of pain intensity in childbirth', *Journal of Clinical Investigation*, 28, 153–162.
40. Tousignant, N.R. (2006). 'Pain and the pursuit of objectivity: Pain-measuring technologies in the United States, c.1890–1975', Doctoral Dissertation, McGill University.
41. Binkley, C.J., Beacham, A., Neace, W., Gregg, R.G., Liem, E.B., & Sessler, D.I. (2009). 'Genetic variations associated with red hair color and fear of dental pain, anxiety regarding dental care and avoidance of dental care', *Journal of the American Dental Association*, 140, 896–905. Manning, E.L., & Fillingim, R.B. (2002). 'The influence of athletic status

and gender on experimental pain responses', *Journal of Pain*, 3, 421–428. Okifuji, A., & Hare, B.D. (2015). 'The association between chronic pain and obesity', *Journal of Pain Research*, 8, 399–408.

42. Pud, D., Golan, Y., & Pesta, R. (2009). 'Hand dominancy – A feature affecting sensitivity to pain', *Neuroscience Letters*, 467, 237–240.
43. https://hellocaremail.com.au/older-people-less-likely-report-pain/.
44. Safikhani, S., & others (2017). 'Response scale selection in adult pain measures: Results from a literature review', *Journal of Patient-Reported Outcomes*, 2, 2–9. Hjermstad, M.J., & others (2011). 'Studies comparing numerical rating scales, verbal rating scales, and visual analogue scales for assessment of pain intensity in adults: A systematic literature review', *Journal of Pain and Symptom Management*, 41, 1073–1093.
45. Cowen, R., Stasiowska, M.K., Laycock, H., & Bantel, C. (2015). 'Assessing pain objectively: The use of physiological markers', *Anaesthesia*, 70, 828–847.
46. Wager, T.D., Atlas, L.Y., Lindquist, M.A., Roy, M., Woo, C.W., & Kross, E. (2013). 'An fMRI-based neurologic signature of physical pain', *New England Journal of Medicine*, 368, 1388–1397.
47. Mead, T. (2011). 'You can't measure pain', *Canadian Family Physician*, 57, 764–764.
48. MacAskill, W. (2015). *Doing Good Better*. London: Gotham Books.
49. They also use variants of QALYs such as the DALY, or disability-adjusted life year.
50. Torrance, G.W. (1987). 'Utility approach to measuring health-related quality of life', *Journal of Chronic Diseases*, 40, 593–600.
51. Whitehead, S.J., & Ali, S. (2010). 'Health outcomes in economic evaluation: The QALY and utilities', *British Medical Bulletin*, 96, 5–21.
52. Pettitt, D.A., & others (2016). 'The limitations of QALY: A literature review', *Journal of Stem Cell Research and Therapy*, 6, 1–7.
53. https://www.cgdev.org/sites/default/files/1427016_file_moral_imperative_cost_effectiveness.pdf.
54. Jamison, D., & others (eds) (2006). *Disease Control Priorities in Developing Countries*, 2nd edn. Oxford University Press. In this case, benefits are measured in DALYs per $1000 of funding.
55. Shedler, J.K., Jonides, J., & Manis, M. (1985). 'Availability: Plausible but questionable', Paper presented at the 26th annual meeting of the Psychonomic Society, Boston, MA. Cited in: Jonides, J., & Jones, C.M. (1992). 'Direct coding for frequency of occurrence', *Journal of Experimental Psychology: Learning, Memory and Cognition*, 18, 368–378.

56. Gigerenzer, G., Todd, P.M., & the ABC Research Group (1999). *Simple Heuristics That Make Us Smart*. Oxford University Press, pp. 219–221.
57. https://www.dailystar.co.uk/news/latest-news/644911/Lottery.
58. https://www.independent.co.uk/voices/commentators/nigel-hawkes-and-our-survey-saysnothing-you-can-rely-on-1920720.html.
59. https://www.express.co.uk/news/uk/947939/tea-cuppa-britons-survey-research-milk-teabag-tetley.

8. It's Probably True: Soft Numbers Meet Hard Data

1. Fisher, R.A., (1955). 'Statistical methods and scientific induction', *Journal of the Royal Statistical Society, Series B*, 17, 69–77.
2. Technically, the probability of getting any exact result in a survey is very low, because a very large number, or even infinite number, of other results could have occurred. For this reason, we ask: 'How probable is it that we would get this result in our survey, *or one that is even further from the hypothesis,* if the hypothesis is true?'
3. To see how Bayes's theorem works, take 100 days when the sky looks similar to today. Your estimate implies that, on days like this, it will rain on 60 days and be fine on 40 days.

   Of the 60 days when it rains, the weather forecast will wrongly indicate fine weather on 10 per cent of occasions. So it will forecast fine weather when it rains on 6 days.

   Of the 40 days when it's fine, the weather forecast will have correctly indicated fine weather on 90 per cent of occasions. So it will forecast fine weather when it turns out to be fine on 36 days.

   This means that on 100 days like today the weather forecast will predict fine weather on 36 + 6 = 42 days. We have one of those days today and we know that when we get this forecast, it only rains on 6 out of 42 days, or on 14 per cent of days.

   A quick way of carrying out the calculation is shown below. Here the probabilities are expressed on a 0 to 1 scale, rather than as percentages, to make the arithmetic easier:

   (1) Complete the following table.

Prior probability the event is true	Probability of getting new information if event is true
1 minus the above probability	Probability of getting new information if event is NOT true

(2) Multiply the two top numbers. Call the answer TOP.
(3) Multiple the two bottom numbers. Call the answer BOTTOM.
(4) Add TOP to BOTTOM. Call this SUM.
(5) The revised (or posterior) probability is TOP / SUM.
So for our example we have:
(1)

0.6	0.1
0.4	0.9

(2) TOP = 0.6 × 0.1 =0.06.
(3) BOTTOM = 0.4 × 0.9 = 0.36.
(4) SUM = 0.06 + 0.36 = 0.42.
(5) The revised probability = 0.06 / 0.42 = 0.14, or 14 per cent.
If you multiply each of the probabilities in steps (2) to (5) by 100, you will see that the numbers correspond to the numbers of days we referred to earlier.

4. See Martyn Hooper's essay at: https://www.york.ac.uk/depts/maths/histstat/price.pdf.
5. https://www.nytimes.com/2014/01/05/magazine/a-speck-in-the-sea.html.
6. Fisher even went as far as to suggest that Bayes must have doubted the soundness of his own ideas because he did not publish them in his lifetime. He went to great lengths to try to establish a method that allowed you to determine a probability of a hypothesis being true while avoiding the need to include subjective judgements. But his approach, which was based on what he called fiducial probabilities (after the Latin for faith or trust), attracted much criticism. This was never resolved in his lifetime, though some researchers to this day are still grappling with the concept to see if it has merit. Note that, less controversially, Bayes's theorem can also be used to update prior probabilities based on objective data when new data becomes available.
7. Robert Matthews, 'Flukes and flaws', *Prospect Magazine*, November 1998, pp. 20–24.
8. Press, S.J., & Tanur, J.M. (2016). *The Subjectivity of Scientists and the Bayesian Approach*. New York: Dover.
9. Fang, F.C., Steen, R.G., & Casadevall, A. (2012). 'Misconduct accounts for the majority of retracted scientific publications', *Proceedings of the National Academy of Sciences*, 109, 17,028–17,033.

10. Press, S. J., & Tanur, J.M. (2016). *The Subjectivity of Scientists and the Bayesian Approach*. New York: Dover.
11. https://www.dispatch.com/article/20100225/NEWS/302259665.
12. Tversky, A., & Kahneman, D. (1974). 'Judgment under uncertainty: Heuristics and biases', *Science*, 185, 1124–1131.
13. There have been many other criticisms of this method for testing hypotheses. See, for example: Ziliak, S.T., & McCloskey, D.N. (2008). *The Cult of Statistical Significance*. Ann Arbor, MI: University of Michigan Press.
14. Over 1.2 million papers were catalogued in 2014 in one academic database and this only contains papers in the higher-quality journals; some estimates suggest that 2.5 million papers are being published around the world each year.
15. Begley, C.G., & Ellis, L.M. (2012). 'Drug development: Raise standards for preclinical cancer research', *Nature*, 483, 531–533.
16. Bem, D.J. (2011). 'Feeling the future: Experimental evidence for anomalous retroactive influences on cognition and affect', *Journal of Personality and Social Psychology*, 100, 407–425.
17. Daniel Engber, 'Daryl Bem proved ESP is real: Which means science is broken', *Slate*, 17 May 2017: https://slate.com/health-and-science/2017/06/daryl-bem-proved-esp-is-real-showed-science-is-broken.html.
18. Open Science Collaboration (2015). 'Estimating the reproducibility of psychological science', *Science*, 349, aac4716.
19. Gervais, W.M., & Norenzayan, A. (2012). 'Analytic thinking promotes religious disbelief', *Science*, 336, 493–496.
20. Camerer, C.F., & others (2018). 'Evaluating the replicability of social science experiments in Nature and Science between 2010 and 2015', *Nature Human Behaviour*, 2, 637–644.
21. Lee, S.W., & Schwarz, N. (2010). 'Washing away postdecisional dissonance', *Science*, 328, 709–709.
22. https://www.newscientist.com/article/2185358-walking-backwards-can-boost-your-short-term-memory/.
23. https://www.newscientist.com/article/2169622-how-your-name-shapes-what-other-people-think-of-your-personality/.
24. Evanschitzky, H., & Armstrong, J.S. (2010). 'Replications of forecasting research', *International Journal of Forecasting*, 26, 4–8.
25. Rotello, C.M., Heit, E., & Dubé, C. (2015). 'When more data steer us

wrong: Replications with the wrong dependent measure perpetuate erroneous conclusions', *Psychonomic Bulletin & Review*, 22, 944–954.
26. Unwin, S. (2004). *The Probability of God: A Simple Calculation That Proves the Ultimate Truth*. New York: Free Rivers Press.
27. Using the table in note 3, we have the following:

Prior probability you are innocent = 999/1000	Probability murderer matches your description = 1/80
Prior probability you are the murderer = 1/1000	Probability murderer matches your description = 1

So TOP = 0.0125; BOTTOM = 0.001; SUM = 0.0135.
Posterior probability you are innocent = 0.0125/0.0135 = 0.926, or about 93 per cent (the difference from the answer in the main text is due to rounding).
28. Angela Saini. 'A formula for justice', *The Guardian*, 2 October 2011.
29. https://understandinguncertainty.org/court-appeal-bans-bayesian-probability-and-sherlock-holmes.

9. To Hell with the Numbers

1. Aylin, P., Best, N., Bottle, A., & Marshall, C. (2003). 'Following Shipman: A pilot system for monitoring mortality rates in primary care', *The Lancet*, 362(9382), 485–491.
2. Spiegelhalter, D., & Best, N. (2004). 'Shipman's statistical legacy', *Significance*, 1, 10–12.
3. Bottle, A., & Aylin, P. (2017). *Statistical Methods for Healthcare Performance Monitoring*. London: CRC Press.
4. Fransen, M.L., Smit, E.G., & Verlegh, P.W. (2015). 'Strategies and motives for resistance to persuasion: An integrative framework', *Frontiers in Psychology*, 6, 1201–1221.
5. Mercier, H., & Sperber, D. (2017). *The Enigma of Reason*. Cambridge, MA: Harvard University Press.
6. Baggini, J. (2017). *A Short History of the Truth*. London: Quercus, p. 100.
7. Gorman, S.E., & Gorman, J.M. (2016). *Denying to the Grave: Why We Ignore the Facts That Will Save Us*. Oxford University Press.
8. Kaplan, J.T., Gimbel, S.I., & Harris, S. (2016). 'Neural correlates of maintaining one's political beliefs in the face of counterevidence', *Scientific Reports*, 6, 39589.

9. Nyhan, B., & Reifler, J. (2010). 'When corrections fail: The persistence of political misperceptions', *Political Behavior*, 32, 303–330.
10. Friesen, J.P., Campbell, T.H., & Kay, A.C. (2015). 'The psychological advantage of unfalsifiability: The appeal of untestable religious and political ideologies', *Journal of Personality and Social Psychology*, 108, 515–529.
11. https://www.lrb.co.uk/v33/n22/jenny-diski/what-might-they-want.
12. Bonaccio, S., & Dalal, R.S. (2006). 'Advice taking and decision-making: An integrative literature review, and implications for the organizational sciences', *Organizational Behavior and Human Decision Processes*, 101, 127–151.
13. Yaniv, I. (2004). 'Receiving other people's advice: Influence and benefit', *Organizational Behavior and Human Decision Processes*, 93, 1–13.
14. Krueger, J.L. (2003). 'Return of the ego – Self-referent information as a filter for social prediction: Comment on Karniol (2003)', *Psychological Review*, 110, 585–590.
15. Yaniv, I., & Kleinberger, E. (2000). 'Advice taking in decision making: Egocentric discounting and reputation formation', *Organizational Behavior and Human Decision Processes*, 83, 260–281.
16. Önkal, D., Goodwin, P., Thomson, M., Gönül, S., & Pollock, A. (2009). 'The relative influence of advice from human experts and statistical methods on forecast adjustments', *Journal of Behavioral Decision Making*, 22, 390–409.
17. Lim, J.S., & O'Connor, M. (1995). 'Judgemental adjustment of initial forecasts: Its effectiveness and biases', *Journal of Behavioral Decision Making*, 8, 149–168.
18. Dietvorst, B.J., Simmons, J.P., & Massey, C. (2015). 'Algorithm aversion: People erroneously avoid algorithms after seeing them err', *Journal of Experimental Psychology: General*, 144, 114–126.
19. Sugiyama, M.S. (2001). 'Narrative theory and function: Why evolution matters', *Philosophy and Literature*, 25, 233–250.
20. John Allen Paulos, 'Stories vs. statistics', *The New York Times*, 24 October 2010.
21. In fact, the more detail we put into a story, the more plausible it is likely to seem, because we are giving reasons for the events. However, more elaborate stories are actually less probable. 'John was the person who shot Michael because he was jealous of his wealth and the way he boasted about it and kept it all to himself' is less probable than 'John was the person who shot Michael'. The latter case allows for the

possibility that John could have shot Michael for any of a large number of reasons.
22. Taleb, N.N. (2007). *The Black Swan: The Impact of the Highly Improbable*. New York: Random House.
23. www.chron.com, 20 March 2017.
24. Singer, P. (2010). *The Life You Can Save*. London: Picador, p. 48.
25. Kogut, T., & Ritov, I. (2005). 'The "identified victim" effect: An identified group, or just a single individual?', *Journal of Behavioral Decision Making*, 18, 157–167.
26. Slovic, P. (2007). '"If I look at the mass I will never act": Psychic numbing and genocide', *Judgment and Decision Making*, 2, 79–95.
27. Slovic, P., Finucane, M.L., Peters, E., & MacGregor, D.G. (2002). 'The affect heuristic', in: T. Gilovich, D. Griffin, & D. Kahneman (eds), *Heuristics and Biases: The Psychology of Intuitive Judgment*, pp. 397–420. New York: Cambridge University Press.
28. *Time*, vol. 193, no. 4–5, 4 February 2019.
29. https://www.robheard.co.uk/the-somme-19240/.
30. Anthony Browne, 'Calf? I nearly died', *The Observer*, 29 April 2001.
31. Small, D.A., Loewenstein, G., & Slovic, P. (2007). 'Sympathy and callousness: The impact of deliberative thought on donations to identifiable and statistical victims', *Organizational Behavior and Human Decision Processes*, 102, 143–153.
32. 'Revealed: Secrets of *The Apprentice*', *Radio Times*, 7 November 2011.
33. Source: Forest Institute of Professional Psychology, Springfield, MO.
34. Patricia Nilsson, *Financial Times*, 12 October 2017.
35. Cassar, G. (2010). 'Are individuals entering self-employment overly optimistic? An empirical test of plans and projections on nascent entrepreneur expectations', *Strategic Management Journal*, 31, 822–840.
36. Flyvbjerg, B. (2008). 'Curbing optimism bias and strategic misrepresentation in planning: reference class forecasting in practice', *European Planning Studies*, 16, 3–21.
37. Weick, M., & Guinote. A. (2010). 'How long will it take? Power biases and time predictions', *Journal of Experimental Social Psychology*, 46, 595–604.
38. Janis, I.L. (1972). *Victims of Groupthink*, Boston, MA: Houghton Mifflin.
39. Goodwin, P. (2017). *Forewarned: A Sceptic's Guide to Prediction*. London: Biteback Publications.

10. Safety in Numbers

1. Mark Lewisohn, *The Independent*, 18 November 2003.
2. Neil Strauss, 'Why we're living in the age of fear', *Rolling Stone*, 6 October 2016.
3. Pinker, S. (2012). *The Better Angels of Our Nature: A History of Violence and Humanity*. New York: Penguin.
4. Uppsala Conflict Data Program: www.ucdp.uu.se.
5. E. De Haro, 'Be not afraid', *The Atlantic*, March 2015.
6. www.bbc.co.uk/news/uk-42182497.
7. Source: World Bank. The figures have been adjusted for inflation.
8. Peter Diamandis: https://singularityhub.com/2017/10/12/why-the-world-is-still-better-than-you-think-new-evidence-for-abundance/.
9. Source: World Bank.
10. https://ourworldindata.org/natural-catastrophes/.
11. Eagleman, D. (2011). *Incognito. The Secret Life of the Brain*. Edinburgh: Canongate.
12. LeDoux, J.E., & Pine, D.S. (2016). 'Using neuroscience to help understand fear and anxiety: A two-system framework', *American Journal of Psychiatry*, 173, 1083–1093.
13. Bonanno, G.A., & Jost, J.T. (2006). 'Conservative shift among high-exposure survivors of the September 11th terrorist attacks', *Basic and Applied Social Psychology*, 28, 311–323.
14. http://www.anorak.co.uk/288298/tabloids/the-daily-mails-list-of-things-that-give-you-cancer-from-a-to-z.html/ (accessed 1 December 2017).
15. Office of Rail and Road. *Rail Safety Statistics: 2015–16 Annual Statistical Release*. There were fatalities on British railways in the eight years up to September 2015, but none of these were passengers who died as a result of a train accident.
16. D. Mosher & S. Gould, 'How likely are foreign terrorists to kill Americans? The odds may surprise you', *Business Insider UK*, 1 February 2017.
17. http://natgeotv.com/ca/human-shark-bait/facts.
18. Myers, D.G. (2001). 'Do we fear the right things?' *APS Observer*, 14(10), 3.
19. Gigerenzer, G. (2006). 'Out of the frying pan into the fire: Behavioral reactions to terrorist attacks', *Risk Analysis*, 26, 347–351.
20. http://news.bbc.co.uk/1/hi/england/london/8236820.stm.
21. http://www.economist.com/node/895855.

22. Julia Hartley-Brewer, 'Neglect of road safety spending "costs lives"', *The Guardian*, 9 February 2000.
23. *Daily Telegraph*, 4 January 2017.
24. https://www.thecut.com/2017/05/wine-alcohol-breast-cancer-risk-study.html.
25. https://www.menshealth.com/health/loneliness-and-heart-attack-risk.
26. *The Independent*, 30 August 2013.
27. Mabry, M.A. (1971). 'The relationship between fluctuations in hemlines and stock market averages from 1921 to 1971', masters thesis, University of Tennessee, Knoxville, TN.
28. http://www.dailymail.co.uk/health/article-5070707/Spanking-children-increases-risk-depression.html.
29. http://www.mirror.co.uk/news/uk-news/netflix-kill-warning-watching-much-8492515.
30. http://www.dailymail.co.uk/health/article-4401442/Having-tattoo-makes-sweat-less.html.
31. http://www.tylervigen.com/spurious-correlations.
32. http://www.nber.org/papers/w18212.pdf.
33. For other limitations of this and other studies, see the excellent *NHS Behind the Headlines* website: https://www.nhs.uk/news/.
34. Myers, D.G. (2001). 'Do we fear the right things?', *APS Observer*, 14(10), 3.
35. Ward, N.J., & Wilde, G.J. (1996). 'Driver approach behaviour at an unprotected railway crossing before and after enhancement of lateral sight distances: An experimental investigation of a risk perception and behavioural compensation hypothesis', *Safety Science*, 22, 63–75.
36. Hendriks, F., Kienhues, D., & Bromme, R. (2016). 'Trust in science and the science of trust', in: Bernd Blöbaum (ed.), *Trust and Communication in a Digitized World*, pp. 143–159. New York: Springer.
37. https://www.newstatesman.com/health/2008/08/asbestos-victims-company.
38. Fox, E., & others (2007). 'Hypersensitivity symptom associated with magnetic field', Mobile Telecommunications and Health Research Programme, Health Protection Agency, Didcot, Oxfordshire, UK.
39. http://www.somersetlive.co.uk/news/somerset-news/vodafones-plans-15-metre-monopole-108614.
40. Sellnow, T.L., Ulmer, R.R., Seeger, M.W., & Littlefield, R. (2008). *Effective Risk Communication: A Message-Centered Approach*. New York: Springer. See also: Stilgoe, J. (2016). 'Scientific advice on the move: The UK mobile

phone risk issue as a public experiment', *Palgrave Communications*, 2, 16028.

41. https://www.theguardian.com/science/head-quarters/2015/jan/16/declinism-is-the-world-actually-getting-worse.
42. Mather, M., & Carstensen, L.L. (2005). 'Aging and motivated cognition: The positivity effect in attention and memory', *Trends in Cognitive Sciences*, 9, 496–502.
43. Soroka, S., & McAdams, S. (2015). 'News, politics, and negativity', *Political Communication*, 32, 1–22.
44. http://news.bbc.co.uk/1/hi/uk/2569781.stm.

11. When Can You Count on a Number?

1. Sasha Harris-Lovett, *Los Angeles Times*, 10 August 2015.
2. Kahneman, D. (2011). *Thinking Fast and Slow*. London: Allen Lane. Note that Kahneman's theory of two mental systems is not without its critics. See, for example: Evans, J.S.B. (2006). 'Dual system theories of cognition: Some issues', In: *Proceedings of the Annual Meeting of the Cognitive Science Society*, vol. 28, no. 28.
3. Shah A.J., & Oppenheimer, D.M. (2007). 'Easy does it: The role of fluency in cue weighting', *Judgment and Decision Making*, 2, 371–379.
4. Rail journey time + Road journey time = 210 minutes.
   But the Rail journey time is the Road journey time + 200 minutes. So, we have:
   Road journey time + 200 + Road journey time = 210 minutes.
   After subtracting 200 from both sides of the equation, we have:
   2 × Road journey time = 10 minutes.
   So: Road journey time = 5 minutes.
5. Bolte, A., Goschke, T., & Kuhl, J. (2003). 'Emotion and intuition: Effects of positive and negative mood on implicit judgements of semantic coherence', *Psychological Science*, 14, 416–421.
6. Alter, A.L., Oppenheimer, D.M., Epley, N., & Eyre, R.N. (2007). 'Overcoming intuition: Metacognitive difficulty activates analytic reasoning', *Journal of Experimental Psychology: General*, 136, 569–576.
7. Chaiken, S. (1980). 'Heuristic versus systematic information processing and the use of source versus message cues in persuasion', *Journal of Personality and Social Psychology*, 39, 752–766.
8. Schwarz, N., Strack, F., Hilton, D., & Naderer, G. (1991). 'Base rates, representativeness, and the logic of conversation: The contextual relevance of "irrelevant" information', *Social Cognition*, 9, 67–84.

9. Crandall, B., & Getchell-Reiter, K. (1993). 'Critical decision method: A technique for eliciting concrete assessment indicators from the intuition of NICU nurses', *Advances in Nursing Science*, 16, 42–51.
10. Biederman, I., & Shiffrar, M.M. (1987). 'Sexing day-old chicks: A case study and expert systems analysis of a difficult perceptual-learning task', *Journal of Experimental Psychology: Learning, Memory, and Cognition*, 13, 640–645.
11. Hodgkinson, G.P., Langan-Fox, J., & Sadler-Smith, E. (2008). 'Intuition: A fundamental bridging construct in the behavioural sciences', *British Journal of Psychology*, 99, 1–27.
12. Kahneman, D., & Klein, G. (2009). 'Conditions for intuitive expertise: A failure to disagree', *American Psychologist*, 64, 515–526.
13. Tetlock, P. (2005). *Expert Political Judgment*. Princeton University Press.
14. Klein, G. (2014). *Seeing What Others Don't*. London: Nicholas Brealey. A Ponzi scheme, named after Charles Ponzi, is a type of fraud in which investors are attracted by the promise of high risk-free returns. Investment returns normally come from sources such as company profits or share price increases. However, in a Ponzi scheme, the funds deposited by those investing later in the scheme are used to provide returns for earlier investors. Therefore, to be sustainable, the scheme requires a constant stream of new investors.
15. Einhorn, H.J., & Hogarth, R.M. (1978). 'Confidence in judgement: Persistence of the illusion of validity', *Psychological Review*, 85, 395–416.
16. Peterson, D.K., & Pitz, G.F. (1988). 'Confidence, uncertainty, and the use of information', *Journal of Experimental Psychology: Learning, Memory, and Cognition*, 14, 85–92.
17. Hall, C.C., Ariss, L., & Todorov, A. (2007). 'The illusion of knowledge: When more information reduces accuracy and increases confidence', *Organizational Behavior and Human Decision Processes*, 103, 277–290.
18. Garcia-Retamero, R., & Cokely, E. T. (2013). 'Communicating health risks with visual aids', *Current Directions in Psychological Science*, 22, 392–399.
19. Tufte, E.R. (1983). *The Visual Display of Quantitative Information*. Cheshire, CT: Graphics Press.
20. Rosling, S., Rönnuld, A.R. (2018). *Factfulness*, London: Sceptre Books.
21. www.dollarstreet.org.
22. Currey, M. (2013). *Daily Rituals: How Great Minds Make Time, Find Inspiration, and Get to Work*. London: Picador.
23. Source: Mintel.

24. https://www.greenpeace.org/international/story/15882/every-minute-of-every-day-the-equivalent-of-one-truckload-of-plastic-enters-the-sea/.
25. Prachi Bhardwaj, *Business Insider*, 19 December 2018.
26. Peters, E., Dieckmann, N.F., Västfjäll, D., Mertz, C.K., Slovic, P., & Hibbard, J.H. (2009). 'Bringing meaning to numbers: The impact of evaluative categories on decisions', *Journal of Experimental Psychology: Applied*, 15, 213–227.
27. Budescu, D.V., Broomell, S., & Por, H.H. (2009). 'Improving communication of uncertainty in the reports of the Intergovernmental Panel on Climate Change', *Psychological Science*, 20, 299–308.
28. Boyle, D. (2000). *The Tyranny of Numbers*. London: HarperCollins.

Appendix. Questions to Ask about a Statistic

1. https://blogs.scientificamerican.com/guest-blog/9-bizarre-and-surprising-insights-from-data-science/.

Index

K

L

M